Praise for *Copernicus' Secret*

"*Copernicus' Secret* at last brings the astronomer to life. . . . No other biography of which I am aware treats the life of this scientific giant more vividly than this one."

—Owen Gingerich, *The New York Times Book Review*

"In his highly readable book . . . Repcheck draws a portrait of an unusual star of the scientific revolution. . . . *Copernicus' Secret* is a lucid biography of a complicated man who paved the way for the scientific revolution against the odds—and somewhat against his will."

—Marty Beckermann, *Discover*

"A fine biography. . . . Repcheck does an exceptional job bringing to life [Copernicus'] character, his era and the astronomical problem he solved. . . . [Copernicus] deserves his place among the founders of modern science, and this lively, lucid account does him justice."

—*Kirkus Reviews*

"Repcheck tells the incredible tale of Georg Rheticus, who made the arduous trip from Lutheran Germany to Catholic Poland to convince Copernicus to release his magnum opus. The tale drives the book, and great narrative scenes, including his betrayal of Rheticus, make Copernicus seem fully human."

—Sam Kean, *New Scientist*

"Excellent. . . . Mr. Repcheck tells [his story] extremely well, and with great flair."

—Eric Ormsby, *The New York Sun*

"Illuminating. . . . The history of science here reclaims a fascinating lost chapter."

—*Booklist* (starred review)

"Far-ranging. . . . Repcheck paints a vivid picture of the times."

—*Publishers Weekly*

"Enjoyable. . . . Repcheck delivers a straight and compelling story, taking us back to a different time and place."

—*Science & Spirit*

Also by Jack Repcheck

The Man Who Found Time:
James Hutton and the Discovery of Earth's Antiquity

Copernicus' Secret

How the Scientific Revolution Began

JACK REPCHECK

Simon & Schuster Paperbacks
New York London Toronto Sydney

Simon & Schuster Paperbacks
A Division of Simon & Schuster, Inc.
1230 Avenue of the Americas
New York, NY 10020

First Simon & Schuster trade paperback edition December 2008

SIMON & SCHUSTER PAPERBACKS and colophon are registered trademarks of Simon & Schuster, Inc.

For information about special discounts for bulk purchases, please contact Simon & Schuster Special Sales at 1-800-456-6798 or business@simonandschuster.com

Designed by Davina Mock-Maniscalco

Maps by Paul J. Pugliese

Manufactured in the United States of America

1 3 5 7 9 10 8 6 4 2

The Library of Congress has cataloged the hardcover edition as follows:

Repcheck, Jack.
Copernicus' secret : how the scientific revolution began / Jack Repcheck.
p. cm.
Includes bibliographical references and index.
1. Copernicus, Nicolaus, 1473–1543. 2. Astronomers—Poland—Biography.
3. Astronomy, Medieval. I. Title.
QB36.C8R387 2007
520.92—dc22
[B]
2007024649
ISBN-13: 978-0-7432-8951-1
ISBN-10: 0-7432-8951-X
ISBN-13: 978-0-7432-8952-8 (pbk)
ISBN-10: 0-7432-8952-8 (pbk)

For Donna,

My wife, my love, my buddy, and my *Angel*—
sharing the Copernicus adventure with you
has made it unforgettable.

Contents

Preface xiii

1. Prelude to Future Troubles 1
2. The Precursors 11
3. Childhood 26
4. Student Years 39
5. Warmia 51
6. Before the Storm 68
7. The Death of the Bishop 81
8. The Mistress and the Frombork Wenches 89
9. The Taint of Heresy 101
10. The Catalyst 109
11. The Nuremberg Cabal 122
12. The Meeting 132
13. The First Summer 140
14. Convincing Copernicus 149
15. The Publication 159
16. The Death of Copernicus 170

17. Rheticus after Copernicus 174

18. The Impact of *On the Revolutions* 181

Notes and Select Sources 197

Suggested Additional Readings 215

Acknowledgments 217

Index 219

Of all discoveries and opinions, none may have exerted a greater effect on the human spirit than the doctrine of Copernicus. The world had scarcely become known as round and complete in itself when it was asked to waive the tremendous privilege of being the center of the universe. Never, perhaps, was a greater demand made on mankind—for by this admission so many things vanished in mist and smoke! What became of Eden, our world of innocence, piety and poetry; the testimony of the senses; the conviction of a poetic-religious faith? No wonder his contemporaries did not wish to let all this go and offered every possible resistance to a doctrine which in its converts authorized and demanded a freedom of view and greatness of thought so far unknown, indeed not even dreamed of.

—Johann Wolfgang von Goethe, 1808

Copernicus' Secret

Preface

A FLAWED AND COMPLEX MAN—distant, obsequious, womanizing, but possessing a profoundly original and daring intellect—started the scientific revolution. His name was Nicolaus Copernicus. He achieved this breakthrough when he published his seminal book *On the Revolutions of the Heavenly Spheres* in 1543, the year that he died, aged seventy. The work provided the technical details for Copernicus's "heliocentric," or sun-centered, theory, the model of the universe that hypothesized that the earth and other planets revolved around the sun, and that the earth itself rotated once a day on its axis.

Prior to the publication of Copernicus's book, the Judeo-Christian world believed that a perfectly still earth rested in the center of God's universe, and that all heavenly bodies—the sun, the other planets, the moon, and even the distant stars—revolved around it. This conviction was based on the teachings of Aristotle and the writings of Claudius Ptolemy. The Church had long embraced the paradigm because it conformed to scripture and placed humans at the center of God's firmament. Copernicus's revolutionary work not only presented an entirely different cosmology, but once accepted, it required a titanic shift in mind-set and belief. No longer the center of God's creation, the earth became just one of the other planets. By extension, the primary position of God's highest creation, humankind, was also diminished.

There were many scholars before Copernicus who cast doubt on the earth-centered ("geocentric") model of the universe, in particular Aristarchas, a contemporary of Aristotle's. Yet, no one until Copernicus attempted to develop a comprehensive and complete *system* to supplant Ptolemy's. This was the key—Copernicus provided all of the data and mathematics that any other serious student of the heavens would need to conduct inquiries using his heliocentric model of the universe.

Though Copernicus's theory had several serious flaws (in particular, his staunchly held belief that all orbits must be perfectly circular), it was fundamentally correct and exhibited the essential characteristics of modern science—it was based on unchanging principles, rigorous observation, and mathematical proof. His contribution was immense. It formed the foundation of future work by Tycho Brahe, Johannes Kepler, Galileo Galilei, Isaac Newton, and finally Albert Einstein. *On the Revolutions*, then, started the scientific journey that has led inexorably to our modern world.

COPERNICUS'S HISTORY-ALTERING BOOK came very close to never being published. After pouring his soul into the manuscript for at least two decades and essentially completing it, the astronomer made no move to finish it or submit it to a publisher, despite strenuous urgings from friends and colleagues in high places. He was not afraid of being declared a heretic, as many assume; rather, he was worried that parts of the theory distilled in the manuscript were simply wrong, or if not wrong, incomplete. Thus, he resolved to keep it a secret.

Then, in the last years of his life, Copernicus became embroiled in two serious and distracting clashes that nearly resulted in the manuscript following Copernicus to his grave, consigned to a trunk among his belongings. One dispute was all too human and typical—it involved a woman who was his mistress. The other was more serious and a product of the times—Copernicus, a cleric in the hierarchy of the Catholic Church, was tainted with the brush of the heretical Lutheran Reformation.

That the manuscript was not buried with its author was the result of a genuinely remarkable turn of events. At the precise moment that Copernicus was most troubled, a young Lutheran mathematics professor from the University of Wittenberg, having made an arduous journey over hundreds of miles of muddy roads, arrived unannounced on his doorstep. Georg Joachim Rheticus, defying a law that banned Lutherans from Copernicus's region, was determined to find the famous but shadowy astronomer and discover whether or not the rumored revolutionary theory of the heavens was true. He was euphoric when he discovered that it was. Rheticus then stayed with Copernicus for most of the next two years to help him complete the manuscript and publish it.

With turmoil swirling around them in the cathedral town of Frombork in northern Poland, the two gifted scientists found peace in each other's company. They worked together to put the final touches on the book that would introduce the heliocentric theory, beginning the era of scientific discovery that eventually led to modern science. But, as with everything involving Copernicus, nothing was simple, and even the straightforward act of publication became a complicated adventure.

This book explores the life of Copernicus, particularly the eventful last twelve years of his life—a dozen years that changed the course of western history.

I HAVE WRITTEN THIS BOOK for the lay reader who knows nothing of the events I describe, except perhaps for having heard of Copernicus and his theory that the earth revolves around the sun. The science I describe is at the simplest possible level. Those readers interested in digging deeper into the science will be directed to additional readings in the Notes and Select Sources and Suggested Additional Readings sections. The goal of this book is to provide a rich, accurate, and especially human account of the events that started the scientific revolution.

1

Prelude to Future Troubles

My lord, Most Reverend Father in Christ, my noble lord:

With due expression of respect and deference, I have received your Most Reverend Lordship's letter. Again you have deigned to write to me with your own hand, conveying an admonition at the outset. In this regard I most humbly ask your Most Reverend Lordship not to overlook the fact that the woman about whom your Most Reverend Lordship writes to me was given in marriage through no plan or action of mine. But this is what happened. Considering that she had once been my faithful servant, with all my energy and zeal I endeavored to persuade them to remain with each other as respectable spouses. I would venture to call on God as my witness in this matter, and they would both admit it if they were interrogated. But she complained that her husband was impotent, a condition which he acknowledged in court as well as outside. Hence my efforts were in vain . . .

However, with reference to the matter, I will admit to your Lordship that when she was recently passing through here from the Królewiec fair with the woman from Elblag who employs her, she

> remained in my house until the next day. But since I realize the bad opinion of me arising therefrom, I shall so order my affairs that nobody will have any proper pretext to suspect evil of me thereafter, especially on account of your Most Reverend Lordship's admonition and exhortation. I want to obey you gladly in all matters, and I should obey you, out of a desire that my services may always be acceptable.

This letter was written by Nicolaus Copernicus, the father of modern astronomy and catalyst of the scientific revolution. For those who picture their icons unblemished, it is jarring. It would be less startling, perhaps, if Copernicus had been a young man when he wrote these lines—even icons are allowed indiscretions when young. But he was fifty-eight years old when he drafted this letter in July of 1531. The background facts of the matter are that Copernicus had a housekeeper (whose identity we do not know) in the 1520s who left his employ to marry. The woman soon separated from her husband (supposedly, as Copernicus says, because he was impotent) and then went to work as a housekeeper for a widow in the city of Elblag, which was about twenty miles southwest of Copernicus's home in Frombork. On one occasion, as the two women were returning from a fair in Królewiec, a major port about forty miles northeast of Frombork, they decided to stop in Copernicus's town and spend the night. The ex-housekeeper stayed at the astronomer's house. Someone reported the visit to Copernicus's superior, who then sent a letter rebuking him. This letter was Copernicus's response.

Several inferences can be made from this note. First, this was not the first time that the correspondent—Bishop Maurice

Ferber—had scolded Copernicus, presumably over a woman ("Again you have deigned to write to me . . . conveying an admonition"). Second, the woman in question remained in close contact with Copernicus even after her marriage ("I endeavored to persuade them to remain with each other"). Third, Copernicus's lengthy explanation suggests that he himself was implicated in causing the break-up of the couple.

Copernicus was a canon in the Church, a cleric who was required to take "first orders," including the vow of celibacy. To become a priest it was necessary to take "higher orders," which was encouraged and which many canons did. Copernicus did not, however. Bishop Maurice Ferber was Copernicus's superior. This incongruous letter is the first hint in the documentary record that the last years of the great astronomer's life would be troubled and complicated, and that he would be immersed in more than just his astronomical studies.

Its incongruity is also emblematic of many aspects of Copernicus's role as the founder of the scientific revolution. He truly defied the odds when he discovered and defined the heliocentric theory.

To begin with, Copernicus was a late bloomer. At a time when young men were sent to university at about age fourteen or fifteen (women were not permitted to enroll in universities in the fifteenth and sixteenth centuries), Copernicus did not begin until he was almost nineteen. The typical student went to university for three years and then started to seek fame and fortune. Copernicus attended his undergraduate university for four years, did not earn a degree, and then studied at three other universities for the next eight years. He spent a total of twelve years as a university student. Today we would call such a student a slacker. In Copernicus's era such a prolonged sojourn was highly unusual.

Second, there is not even a whiff of ambition emanating from Copernicus's life, nothing to indicate that he might pursue a line of inquiry that would be revolutionary. Most of the other titans of the scientific revolution—certainly Leonardo, Brahe, Galileo, and Newton—had ambitious streaks and outsized egos, and each was eager for acclaim and recognition. Not Copernicus. He was a retiring hermitlike scholar who wanted nothing more than to be left alone. After finally finishing his languid studies, Copernicus followed the path of least resistance and took a position as the personal assistant to his uncle, Lucas Watzenrode, who was an influential prince-bishop in the Church. Watzenrode wanted to groom Nicolaus to be his successor, thus guaranteeing a life of riches and power. Instead, Copernicus opted off the fast track at the age of thirty-seven, left his uncle's side, and spent the rest of his thirty-three years as a comfortable, but minor, cleric. He did not even bother to take the relatively easy step necessary to become a priest, ignoring pressure to do so from his superiors and friends.

Third, and perhaps the most surprising, Copernicus was not even a professional astronomer. Today, amateur astronomers are numerous, and occasionally one will discover a new star or another body in the night sky. But most of the truly important work in the field occurs at universities and observatories, where the best and brightest can become absorbed in their studies, use the most advanced equipment, and interact with other gifted scientists, enriching one another's research and helping one another with difficult calculations. Though the system was much less formalized five hundred years ago when Copernicus was starting, it was even then the case that truly gifted astronomers and mathematicians were identified early and either made professors at universities or else appointed court astronomers/astrologers at the palaces of royalty, nobility, or the clergy. For instance, the most famous astronomer before Copernicus, known as Regiomontanus, was identified as a science prodigy while still a teenager, was quickly made a university professor, and then was supported in turn by a prominent cardinal, the king of Hungary, and finally one of the richest men in Europe. Similarly, the most famous astronomers to follow Copernicus—Brahe, Kepler, and Galileo—all had very rich, powerful benefactors (the king of Denmark, the Holy Roman Emperor, and the Medici, respectively). As professionals, they could devote all their energies to their observations, studies, and writings. By comparison, Copernicus had many distracting official duties as a canon in the Church, and he practiced astronomy only as an avocation.

The list goes on and on. Copernicus worked with inferior, primitive instruments—not even close to the state of the art at the time—and he lived more than half a century before the invention of the telescope. He resided in an inhospitable place for observing the night sky—northern Europe near the Baltic Sea. He worked alone most of the time—astronomy has nearly always been practiced with others (keeping each other awake in the middle of the night, helping to lift and maneuver large instruments, confirming sightings, etc.). Finally, it even appears that he miscalculated precisely where he lived, which caused many of his observations to be inaccurate. In sum, Copernicus should have stood no chance of making even a minor mark in astronomy, let alone one of the most fundamental discoveries in the history of the science.

ALTHOUGH COPERNICUS was the unlikely vessel, the *era* in which he lived was ripe for an intellectual revolution in astronomy. A hundred years after the Black Death had annihilated one-third of the population of Europe, the Renaissance began in the second half of the fifteenth century. A spirit of renewed vigor infused all the intellectual disciplines, from poetry to engineering. Astronomy was among the fields reignited. Christopher Columbus's discovery of the New World in 1492 kicked the Renaissance into high gear. At the time Copernicus was an undergraduate at the University of Krakow, in the Polish capital. Then, with the Renaissance in full flower—Leonardo began painting the *Last Supper* in 1495, Michelangelo sculpted the *Pietà* the same year, da Gama quickly followed in Columbus's footsteps by sailing around the Cape of Good Hope and landing in Calcutta in 1498—Copernicus became a graduate student in the center of the intellectual ferment, Italy, where he studied at the universities of Bologna, Padua, and Ferrara for the next eight years.

The idea that Europeans were living at a momentous time was in the air in the late 1490s, and Copernicus breathed it greedily. And then, *he* became a participant. Sometime in late 1496 or early 1497, Copernicus started renting rooms from, and then assisting, one of the leading astronomers of the day, even while he was ostensibly studying canon law at the University of Bologna. Domenico Maria de Novara was not only an astronomy professor at the university, he was also Bologna's chief astrologer. He was charged with producing predictions about the future health and well being of the nobility and clergy of the city. Novara was representative of the way astronomy was practiced during Copernicus's era—the formal study of the heavens and the "reading" of what the movements in the heavens meant to people on Earth were inseparably linked. To be an astronomer was to be an astrologer and vice versa.

Working with Novara, Copernicus made his first known formal observation in March 1497, when he witnessed the lunar eclipse of the bright star Alpha Tauri. Novara's numerous writings make it clear that he was skeptical about parts of the prevailing Ptolemaic (Earth-centered) system, and Copernicus and he no doubt discussed these conundrums. During this period of his life Copernicus embraced astronomy wholeheartedly.

But if Copernicus was astonishing those around him with his math and astronomy skills in the late 1490s and early 1500s, while he was still in Italy and in his prime intellectual years, some mentor should have latched on to him and brought him into the astronomy/astrology "club" as either a court astrologer or university professor. The fact that this did not happen could be due to several factors. Perhaps he had not impressed the astronomy community in Italy. Or perhaps he did not have a strong enough belief in astrology to make it part of his life's work, even if this meant having the freedom to devote most of his energies to scientific work. Or he may have felt so beholden to his uncle Lucas, who had been his benefactor for years, that he was compelled to begin the life course that the bishop had set for him. In all likelihood, Copernicus's departure from Italy in 1503 was the result of some combination of these reasons.

In that year, Nicolaus, now thirty years old, completed his studies with a doctorate in canon law from the University of Ferrara, and returned to Poland. There he began his career in the Church hierarchy. But even that pursuit was stillborn. As mentioned earlier, Copernicus's powerful uncle chose him to be his assistant, and expected him to one day take over his seat as the prince-bishop of Warmia, a principality in northern Poland. But Copernicus was not interested. After seven years in the service of his uncle, he resigned as the bishop's secretary. In 1510, he took up the official post from which he actually received his income from the Church, as a canon at the cathedral in Frombork, on the Baltic coast and not far from the port city of Gdansk.

Though he did not make an impact on the astronomers in Italy a decade earlier, they must have missed indications of his talents. Because here in little, inconsequential Warmia, which Copernicus himself described as "the remotest corner of the world," far away from the universities in Kraków, Bologna, Padua, and Ferrara, something extraordinary happened. Over the next few years and completely on his own, Nicolaus Copernicus formulated one of the greatest achievements in the history of human thought.

Copernicus the astronomer, not the nephew of the bishop, emerged from nowhere sometime before 1514. It was about then that he penned his astronomical ideas in a short, anonymous, untitled, non-technical, handwritten manuscript. He made copies of this document and sent them to interested friends. The recipients were probably former fellow students from his university days in Kraków and Italy.

The document was essentially a manifesto for the heliocentric theory. Copernicus began by announcing that the Ptolemaic system "presents no small difficulties" and that it had "defects." In writing these words he was directly confronting well over thirteen hundred years of firm belief and rigorous support. He went on to say, "I often considered whether there could perhaps be found a more reasonable arrangement . . . in which everything would move uniformly about its proper center, as the rule of absolute motion requires. After I had addressed myself to this very difficult and almost insoluble problem, the suggestion at length came to me how it could be solved with fewer and much simpler constructions than formerly used, if some assumptions were granted to me." He then proceeded to list seven assumptions or axioms. The axioms were stunning in their novelty, and several generations later would be viewed as heretical. Three were particularly astonishing: "The center of the earth is not the center of the universe, but only of gravity and of the lunar sphere" (meaning that the moon revolves around the earth, but only the moon); "All the spheres revolve about the sun as their midpoint, and therefore the sun is the center of the universe"; and "The earth . . . performs a complete rotation on its fixed poles in a daily motion." So there on page two of the essay was the first utterance of the theory that started modern science— ". . . therefore the sun is the center of the universe."

Immediately after presenting his assumptions, Copernicus stated, "I have thought it well, for the sake of brevity, to omit from this sketch the mathematical demonstrations, reserving these for my larger work."

The *Commentariolus*, as the essay later came to be called, circulated throughout the astronomy community of Europe as the original recipients made copies to send to their colleagues and then filed away their own copies. Through this network the heliocentric theory became known. A few knew who the author was, most did not, and at least some in the astronomer-astrologer community eagerly awaited the promised "larger work." And waited, and waited.

A few years passed, then the decades of the 1520s and 1530s, and still no larger work containing the needed proof of Copernicus's startling assertions appeared. Most readers of the *Commentariolus* probably dismissed the original essay as a fascinating but naïve effort that remained unproven.

But true to his word, Copernicus did produce a deeply learned and highly technical manuscript that confirmed the key tenets of the heliocentric theory. Although he toiled over it for more than twenty years, he made no effort to have the extraordinary work published. Nor does it appear that he showed it to more than a handful of people. He would later write in his masterwork that he was afraid that "when I attribute certain motions to the terrestrial globe, they [he does not say who "they" are] will immediately shout to have me and my opinion hooted off the stage." Copernicus continued, "Therefore, when I weighed these things in my mind, the scorn which I had to fear on account of the newness and absurdity of my opinion almost drove me to abandon a work already undertaken." In fact, "I had kept [the manuscript] hidden among my things."

THE "LARGER WORK" might have been abandoned and never have appeared in print if not for the events about to be described. Perhaps *On the Revolutions of the Heavenly Spheres,* as the larger work was eventually called, would have been published posthumously by one of Copernicus's few friends who knew about it, but its technical nature would have made that difficult. What is certain, though, is that the beginning of the scientific revolution would have been different and surely delayed if the manuscript for *On the Revolutions* had not been published how and when it was.

HOWEVER, before explaining how Nicolaus Copernicus was finally persuaded to release his comprehensive, one-of-a-kind manuscript for publication, it is necessary to first explore how the manuscript—and specifically the ideas contained in it—came to be in the first place. And for that it is necessary to begin twenty years before Copernicus's birth, in a classroom in Vienna.

2

THE PRECURSORS

THE PUBLICATION of Copernicus's manuscript started the scientific revolution in the spring of 1543. However, the first step on the road to that seminal event was taken in a wood-paneled classroom at the University of Vienna nearly a century earlier, in the spring semester of 1454. The course taught in that cramped room in an old building surveyed the latest thinking in theoretical astronomy. It brought together an inventive humanities professor named Georg Peurbach (1423–1461) and an eighteen-year-old student who would be galvanized by the lectures and would quickly become the professor's research and observation partner. The student was Johannes Müller (1436–1476). He would eventually be known by the Latin name Regiomontanus (which means "king's mountain," a reference to his birthplace) and would soon surpass his mentor as the greatest astronomer of the fifteenth century.

The timing of Peurbach's astronomy class could not have been more appropriate. Though neither instructor Peurbach nor pupil Regiomontanus realized it at the time, the Renaissance began at essentially the same time as their partnership. The fall of Constantinople to the Ottoman Turks in 1453 and the invention of the printing press by Johann Gutenberg in Mainz a few years later were two of the key catalysts of the Renaissance. Though Peurbach was just trying to give clear and careful lectures on a knotty subject, and Regiomontanus was just trying to take accurate notes, the intellectual awakening that started in that Viennese classroom also contributed dramatically to the Renaissance.

In fact, if not for their untimely deaths at the ages of thirty-eight and forty, respectively, either Peurbach or Regiomontanus might have become the father of modern astronomy and the scientific revolution instead of the canon of Frombork. In many ways, the two Viennese astronomers played the role of Christopher Marlowe to Copernicus's Shakespeare in the history of modern astronomy—that is, they advanced astronomical scholarship as no one had before them, and they created an environment for an innovator like Copernicus. Their chief joint contribution was to recognize the need for the dramatic reform of astronomy. Yet, before their deaths neither attempted a complete rethinking of the Ptolemaic system. That bold breakthrough would have to wait for the intellectually fearless Copernicus.

GEORG PEURBACH was not even supposed to ponder astronomy, let alone teach a course on the topic. He was a humanities professor, hired by the University of Vienna in 1453 to teach courses on Horace, the *Aeneid*, rhetoric, and other related topics. The young intellectual also wrote poetry in Latin, some of the best from this era. Yet, beginning in 1451, when he recorded his first observations of the heavens, Peurbach became fascinated with astronomy and astrology. He convinced the university rector to allow him to teach a course on astronomy just one year after his appointment as professor. There is no record of Peurbach ever taking a formal course on the subject, so it appears that he was self-taught.

By closely studying the standard textbook of the period,

which was *A Theory of the Planets* (*Theorica planetarium*) and was believed to have been written by one Gerard of Cremona two centuries earlier, Peurbach knew that there were defects with the prevailing model of how the bodies in the night sky moved. He was eager to explore these flaws carefully and perhaps find a remedy for them. Using the forum of a course that lasted several months, he hoped to begin the process.

Gerard of Cremona's text was based directly on the source for all astronomical thinking of the time, Claudius Ptolemy's *Almagest*. Ptolemy (AD ca. 85–165) was a Greek who lived in the Egyptian city of Alexandria. He was a scholar of prodigious talent—incredibly ingenius and prolific. He wrote several magisterial works, the most significant being the *Almagest* (ca. AD 150), which surveyed everything then known about the universe and the study of it. It included tables for locating the heavenly bodies, which was what astrologers coveted most.

The arrangement described within the pages of the *Almagest* was based on Aristotle's conception of the cosmos. The earth was at the center. It did not move at all. It was believed that if the earth *did* move, the atmosphere itself would blow away, and since it clearly was not blowing away, just as clearly the earth must be perfectly still. All of the planets, which included the moon and the sun, revolved around the earth, and they did so in perfect circles, at constant speed, and in the same plane. The planet closest to the earth was the moon, next closest was Mercury, followed by Venus and the sun, then Mars, Jupiter, and finally Saturn (Neptune and Uranus were not discovered until after the invention of the telescope).

The sun and moon were considered unique, but the five other planets were called the "wandering stars," because they moved through the night sky, or wandered (the word "planet" is derived from the Greek word for wanderer). The wandering stars were viewed against the "fixed stars," which were all of the other illuminated objects. The fixed stars also moved around the earth, but they always maintained their position relative to one another. So it was believed that the fixed stars were attached to a celestial sphere that marked the outer boundary of the universe, just beyond Saturn, and this sphere rotated around the earth on a celestial axis once a day. The wandering stars were always seen in a thin band on the dark horizon, which meant that only a few constellations of fixed stars served as their background. The twelve special constellations were called the houses of the zodiac (which means circle or circuit).

Ptolemy's model of the universe was founded on Aristotle's dictates, coupled with what the Alexandrian knew about the actual behavior of the heavenly bodies based on years of careful observation. A troubling conundrum that he had to explain was "retrograde motion"—the concept that the wandering stars, during their annual rotation around the earth, appeared to stop and then actually go in reverse before stopping again and then resuming their proper course. Ptolemy cleverly solved this mystery. In his conception, the planets revolved around the earth by being attached to one of two spheres. Each planet had two spheres. The main sphere, the one that had the earth at its center, was called the "deferent." The second, smaller sphere, to which the planet was attached, was called the "epicycle." The epicycle revolved around a point on the deferent. So the construct was a sphere whose center was on the edge of a much larger sphere: Picture (in two dimensions) a tambourine and one of the round cymbals attached to it; the perimeter of the tambourine is the deferent, and the small cymbal is the epicycle. By the fifteenth century, astronomers believed that Ptolemy's spheres were crystalline—that is, actual, tangible, glasslike spheres.

There were two other complications. First, Ptolemy's observations made him sure that the earth was *not quite* the precise center of the universe. So, he created a point near the earth called the "eccentric." The eccentric *was* the center of all of the planetary deferents. Second, the planets did not move at a consistent, uniform speed, either. This problem was solved by adding a final piece to the puzzle. On the other side of the eccentric, exactly opposite the earth and at the same distance from the eccentric was another point called the "equant." The equant (that is, equalizing point) was the point around which the planet revolved at uniform speed.

So, the deferent and the epicycle addressed the problem of retrograde motion. And the eccentric and equant addressed the problems of locating the true center of revolution and nonuniform speed. But a planet revolved around the eccentric at uniform distance, not at uniform speed. Likewise, a planet moved with uniform speed around the equant, but not at uniform distance. One scholar observed that "Ptolemy broke sharply away from the previous requirement that circular motion must be uniform around its own center." This break from "previous requirement" would deeply trouble Copernicus.

All of these qualifications in the Ptolemaic, geocentric model caused King Alfonso X of Castile, the sponsor of the *Alfonsine Tables,* which were compiled in the thirteenth century and would be the standard source for locating planet and star positions during Copernicus's time, to quip, probably apocryphally, "If the Lord Almighty had consulted me before embarking upon creation, I should have recommended something simpler."

Yet, though Ptolemy's construction was significantly more complex than Aristotle's tidy universe where the earth resided in the middle of a bunch of nested circles, it worked amazingly well. The wandering stars were usually where Ptolemy said they would be, and eclipses and other phenomena occurred reasonably close to when they were expected. Still, by the fifteenth century the model was getting creaky, and discrepancies in the prediction and actual occurrence of eclipses and other astronomical phenomena were getting more pronounced.

SITTING IN PEURBACH'S CLASSROOM was the brightest genius of the fifteenth century, even more brilliant than Leonardo da Vinci, who appeared a generation later. Regiomontanus, born in 1436, was a genuine child prodigy. The son of a miller, he was sent to the University of Leipzig in 1447, at the age of eleven. The boy studied at Leipzig for three years, after which he enrolled at the University of Vienna. The precocious Regiomontanus was already constructing complicated astronomical tables even before arriving at Vienna in 1450, so he jumped at the chance to take Peurbach's new class. Regiomontanus's astute observations and questions led Peurbach to quickly surmise that he had more than a gifted student sitting before him—he had a peer. From this point on, they became colleagues and partners in astronomical observations, despite the fact that Regiomontanus was thirteen years younger than Peurbach.

The next few years were a whirlwind of activity. It almost seems as if Peurbach had a premonition of his impending death, because he accomplished a tremendous amount in a very short period. He and Regiomontanus made a number of accurate astronomical observations together. Peurbach built dozens of well-crafted instruments, such as astrolabes and sundials, and he formulated innovative trigonometric tables. Most importantly, he wrote two of the most important works of the fifteenth century. First, he refined his lecture notes from his 1454 course to produce the *New Theory of the Planets* in manuscript form. The actual innovations in Peurbach's text were minor. However, by correcting certain errors and simplifying parts of Ptolemy's model, Peurbach's text marked a significant improvement over Gerard of Cremona's, which was overly complicated and riddled with mistakes.

Second, he created the *Tables of Eclipses* (*Tabulae eclipsium*, completed in 1459), which accurately projected lunar and solar eclipses for many decades ahead, and became a "must have" reference for astronomers and astrologers for generations. The *Tables* consisted of approximately one hundred dense pages of careful calculations—future astronomers and astrologers were saved hours of painful mathematics because of them.

While enthusiastically fulfilling his duties as humanities professor by teaching courses and mentoring students each semester, Peurbach also wrote annual astronomical yearbooks, which projected the movements of the sun, moon, and planets for the given year. This knowledge was critically important for the landowners and peasants who worked their land (When were the solstices and full moons?), calendar makers (When was Easter?) and astrologers. And Peurbach found time for his own astrology, too. He served as the court astrologer to King Ladislaus V of Hungary, and then the ruler of the Holy Roman Empire, Frederick III. An example of the kind of predictions, or "prognostications," he performed occurred in 1456 when he and Regiomontanus observed what was later named Halley's comet; Peurbach wrote that it signaled drought, pestilence, and war for Greece, Italy, and Spain, and that it spelled trouble for individuals whose nativities have Taurus in the ascendant.

Peurbach also worked on a third key manuscript, but this one he would not see through to completion—an abridged yet authoritative Latin translation (called an "epitome") of Ptolemy's *Almagest*. In the spring of 1460, Cardinal Johannes

Bessarion (1403–1474), the pope's representative to the Holy Roman Empire and a noted Greek scholar, came to Vienna on official Vatican business (he was trying to drum up support for a crusade against the Turks to oust them from Constantinople), but he also found time to seek out the renowned Peurbach and urge him to write a new Latin translation of Ptolemy's landmark book. Bessarion was a native Greek speaker who had left Constantinople ahead of the Ottoman Turks. He was committed to preserving and distributing the Greek classics, many of which he had personally brought with him from the Byzantine capital, which meant arranging for authoritative Latin translations. Bessarion offered financial support for the endeavor from the coffers of the Church. Peurbach enthusiastically accepted the challenge.

Unfortunately, one year later and about halfway through the project, Peurbach became fatally ill. On his deathbed he asked his star student and now colleague to continue the important work on the *Almagest*. As related by Regiomontanus: "When my teacher was dying . . . he commanded me to do these [chapters] as quickly as possible. . . . He was anxious even on his deathbed to fulfill [his duties on the *Almagest*]." How could Regiomontanus say no? Secure in the knowledge that his last effort would be completed, Georg Peurbach, one of the first renaissance men of the Renaissance, passed away at the age of thirty-eight in April 1461.

AS PEURBACH'S SUCCESSOR on the *Almagest*, Regiomontanus became heir to the patronage of Bessarion. He left the University of Vienna about six months after Peurbach's death to accompany the cardinal back to Rome. Regiomontanus would spend the next four years at Bessarion's side. The cardinal helped Regiomontanus to master Greek. He now possessed four unusual talents—he was a gifted mathematician, a rigorous astronomer, an innovative instrument maker, and he could speak and read Greek. This was an unprecedented combination.

Regiomontanus probably finished the *Epitome of the Almagest* in 1462, yet it would not be printed and published until 1496. This was a critically important book, for it not only made Ptolemy's work more widely available, but Regiomontanus also augmented sections with commentary and new observations.

In 1464, Regiomontanus completed a second important work, this one picking up where Peurbach had left off on his trigonometric studies and entitled *On Triangles of Every Kind* (*De triangulis omnimodis*). In the fifteenth century, trigonometry was primarily the study of triangles, and triangulation was an important tool for measuring the position of heavenly bodies, similar to surveying on the surface of the earth. The introduction provided a glimpse of Regiomontanus's personality:

> You, who wish to study great and wondrous things, who wonder about the movement of the stars, must read these theorems about triangles . . . For no one can bypass the science of triangles and reach a satisfying knowledge of the stars . . . A new student should neither be frightened nor despair. Good things are worthy of their difficulties . . . Nor should anyone avoid this small volume. It is a foundation for so many excellent things. I do it no injury by calling it the foot of the ladder to the stars.

His celebrity growing, in 1467 Regiomontanus accepted an invitation from King Matthias Corvinus of Hungary to take up residence in Buda and join the staff of the Royal Library of Hungary. Regiomontanus quickly added to his fame when, shortly after his arrival in Buda, the king fell seriously ill. Members of his court despaired, but Regiomontanus checked his astrological tables and declared that the king was fit and merely suffering from a temporary weakening of the heart caused by a recent eclipse; the king would be healthy in no time. Matthias did in fact recover shortly afterward. His assessment of the king's health reflects how Regiomontanus consistently coupled his astronomical work with astrological work, as did every other astronomer of his era. He wrote short pieces devoted to better understanding the houses of the zodiac and the best times for doctors to practice blood letting according to the position of the moon. He also cast horoscopes for his patrons.

While in Hungary, Regiomontanus completed several more works in trigonometry that were actively used in the early sixteenth century. The most important was titled *The Tables of Directions,* which was printed in 1490. The *Tables* helped to determine the positions of heavenly bodies based on the perceived daily rotation of the night sky; the work was used by astrologers trying to determine the houses of the zodiac. It would be one of the first books owned by Copernicus.

Skeptical of the veracity of Ptolemy's model of the heavens after finishing the *Epitome,* Regiomontanus by 1470 was convinced that an entirely new astronomical system was needed. But he did not know what that system should look like. He wanted to begin at the beginning, which meant obtaining new and completely dependable observations. Buda became a less-than-hospitable location when King Matthias precipitated a war with Bohemia, so Regiomontanus looked elsewhere to continue his research.

In 1471, Regiomontanus packed up his instruments, manuscripts, and other personal belongings and moved to Nuremberg. He explained his reasons to a correspondent:

> Quite recently I have made observations in the city of Nuremberg . . . for I have chosen it as my permanent home not only on account of the availability of instruments, particularly the astronomical instruments on which the entire science is based, but also on account of the great ease of all sorts of communication with learned men living everywhere, since this place is regarded as the center of Europe because of the journeys of the merchants.

Nuremberg remained a vibrant trading center in Copernicus's day and would later play a pivotal role in the publication of *On the Revolutions*.

Once there, Regiomontanus was befriended by a rich merchant named Bernard Walther (1430–1504), who became his patron and student. Walther paid for the construction of one of the first formal astronomical observatories in Europe. It was built outside the protective walls, to the northeast and on a rise above the city. Upon its completion, the two set out on one of the most ambitious astronomical observation programs in history, and certainly the boldest ever undertaken up to that point.

Regiomontanus's plan for reforming astronomy consisted of two prongs. The first was new observations. The second was the publication of important works in mathematics and astronomy. As with the observatory, Walther provided the funds necessary for the now legendary astronomer to start the first publishing operation devoted to the production of mathematical and scientific works.

Regiomontanus set up the printing press in his own house. The first book that he published was Peurbach's *New Theory of the Planets*, which he printed and began selling in 1472. It was a profoundly successful text and probably was the most used book on theoretical astronomy in the sixteenth century (it went through at least fifty-six printings and was translated into several languages, in addition to the original Latin, over nearly two hundred years).

His second published work was one that he authored himself (and was the most popular of his works), the *Ephemerides*, which appeared in 1474 and which accurately projected the positions for the celestial bodies for *every* day from January 1, 1475, to December 31, 1506. The main purpose of the book was to assist astrological readings. The book became even more legendary when it was later reported that Columbus, who traveled with a copy on his voyages, was able to frighten the hostile natives on Jamaica by relying on Regiomontanus to predict a lunar eclipse (February 19, 1504). Apparently the Jamaicans were not allowing Columbus and his men to gather food and water, and when Columbus referred to his copy of the *Ephemerides* and realized that a lunar eclipse was about to occur, he told the Jamaicans that God was going to punish them for threatening his men by taking the moon away. The eclipse occurred exactly as Regiomontanus had predicted, and the natives were awed.

Regiomontanus was thus the first publisher dedicated to rigorous books in the sciences. He even produced a broadsheet called the "Index of Books" that was circulated throughout Nuremberg and then the rest of Europe. The purpose of the Index was to seek individuals who might have copies of the works he wished to publish. He listed more than twenty classics by the ancients (mainly the Greeks, including Ptolemy and

Euclid) that he planned to translate into Latin and then publish. He then described twenty-one projects that he himself was going to write.

Not only did Regiomontanus want to get important works into print, he was also just as serious about having them published well. He recognized that there were unique challenges to publishing mathematical works—these were difficult books to set because of the need to include numerous diagrams and often elaborate tables. Not only must the pages be clear and legible, they must also be free of error. So accuracy became paramount, yet these were precisely the kinds of titles inherently prone to errors:

> For if I am not mistaken we are sinning when we obscure the opinions of noble authors by contaminating them with our own ignorance and infecting posterity with erroneous copies of books. For who does not realize that the admirable art of printing recently devised by our countrymen is as harmful to men if it multiplies erroneous works as it is useful when it publishes properly corrected editions?

Perhaps because Regiomontanus was so careful, he did not succeed in getting many works published before his death in 1476—only four of the forty-one books cited in the Index.

Regiomontanus was a difficult individual and by now he had a huge ego. He had no hesitation in identifying inferior scholars in print. In his Index, when referring to his desire to publish a new translation of *Cosmographia* (another important book by Ptolemy), Regiomontanus wrote, "[T]hat old translation of the Florentine James Angel is vicious. Although he meant well, he was quite weak in his knowledge of both Greek and mathematics." And of George Trebizond, the astronomer-publisher, he wrote, "[A]nyone will see how superficial was his commentary on the *Almagest* and how poor his translation of that work of Ptolemy." Earlier he had said of Trebizond, "You are the most impudently perverse blabbermouth!"

In 1475, Regiomontanus was summoned to Rome by Pope Sixtus IV himself to assist in the reform of the calendar. One of the most important uses of astronomy was for creating accurate calendars. By the late fifteenth century, everyone knew that the calendar in use (the Julian calendar) had grown terribly inaccurate, and it was becoming embarrassing for the Church as it sought to identify the proper Sunday for Easter and other holy days (the Church insisted on observing Easter on the first Sunday following the first full moon after the spring equinox, so astronomical precision was important). Regiomontanus left in the summer, and he died one year later, on July 6, 1476, while still in Rome. He was only forty years old and had left Nuremberg in perfect health. There are two theories about his death. The first and most likely is that he died from the plague, which recurred that year in and around Rome. The second is more sensational—and would make the comparison to Christopher Marlowe even more apt. He was rumored to have been murdered by the two sons of George Trebizond, whom Regiomontanus had so scathingly criticized in print for being a flawed mathematician and astronomer. Either way, it was a horrible end for one of the greatest minds in history.

All told, Regiomontanus wrote four books that became valued standard works well into Copernicus's time and beyond: The *Epitome of Ptolemy's Almagest, On Triangles of Every Kind,* The *Tables of Directions,* and The *Ephemerides.*

* * *

When Regiomontanus died, the astrology and astronomy community knew that it had lost a Promethean figure. Still, there was reason to be optimistic that several talented successors would quickly emerge after his passing. After all, the important books that he published and the seminal manuscripts that he wrote and left behind created a firm foundation for the next generation of astronomers—a foundation that Peurbach and Regiomontanus did not have when they came of age. The universities of Europe were thriving, many of them teaching courses based on Peurbach's and Regiomontanus's manuscripts and books. And Nuremberg had an observatory now.

Yet, no one did emerge. Competent scholars existed at the universities, but there were no transcendent geniuses. In Nuremberg, the logical place for the next great astronomer to surface, Regiomontanus's legacy was celebrated and preserved by a talented group of scholars. Though they continued to do solid work, they were not destined to wear Regiomontanus's mantle.

The gifted individual who would succeed Regiomontanus was but an infant when Regiomontanus was publishing his important books in Nuremberg, and just a toddler playing with his brother and sisters when Regiomontanus passed away in Rome.

3

Childhood

When Regiomontanus was in the middle of his brief residence in Nuremberg, and at approximately the same time that he published Peurbach's *New Theory of the Planets,* three hundred miles to the northeast in the Hanseatic League town of Toruń, Mikolaj Kopernik was born on February 19, 1473. Kopernik would be known to posterity by his latinized name, Nicolaus Copernicus. He would eventually succeed where Regiomontanus had only dreamed—for Copernicus would correctly revise the prevailing theory of the night sky.

Nicolaus Copernicus was born, lived his childhood, and then spent most of his adult life in a very unusual region. The Baltic Littoral of northwestern Poland was, in Copernicus's time, known as Prussia. It was the territory roughly traced by the triangle formed by the major cities of Gdansk to the northwest, Toruń to the southwest, and Królewiec to the northeast as the focal points (see map on pp. 28–29). In between these sizable cities were gently rolling hills filled with forests, fertile farmland, streams, ponds, and countless small towns and villages. The Baltic Sea was its northern border.

This territory has been fought over many times in its history. During the last phase of World War II, Soviet tanks relentlessly pursued the fleeing Germans, who were retreating from the Eastern Front. The area had previously been occupied by the Swedes and by Napoleon's forces. Yet it still bears the unmistakable mark of the people who built the first permanent towns and structures in the region—the Teutonic Knights.

The Teutonic Knights were formed in 1198 in Acre for Holy Roman Emperor Henry VI's crusade to the Holy Land. Officially called the "Order of the Knights of the Cross," they were one of the three special armies formed during the Crusades, the other two being the Knights Templar and the Hospitalers. With the Crusades over, in 1226, Conrad of Mazovia, a duke of Poland, invited the grand master of the Teutonic Order, Hermann von Salza, to move his men into the Baltic region in order to tame and Christianize the pagan natives. Known as the Prussians, they were among the last non-Christians in Europe and were considered a menace by the Christian Poles who shared a border with them. The Knights were promised the land that they conquered.

The warriors viciously took control of the entire area along the northern tier of Poland, a much larger territory than Conrad of Mazovia had agreed to. They slaughtered or enslaved the Prussians, and seized all the land. Within several generations, the extinction of the native Prussians was complete.

The Knights invited fellow Germans and Dutch to move east and settle in Prussia. Thus they established their own kingdom, the Teutonic Order State of Prussia, separate from the Holy Roman Empire to the west, the Kingdom of Poland to the south, and the Kingdom of Lithuania to the east.

The land was bountiful and the numerous rivers allowed for brisk trade within and without the Teutonic Knights' territory. Plus, their new kingdom was blessed with another endowment. The Baltic was the only area then known where the gem amber could be found. (Amber is fossilized tree resin, and has always been coveted for its rarity and presumed medicinal qualities. It was actively traded all over Europe at least as far back as

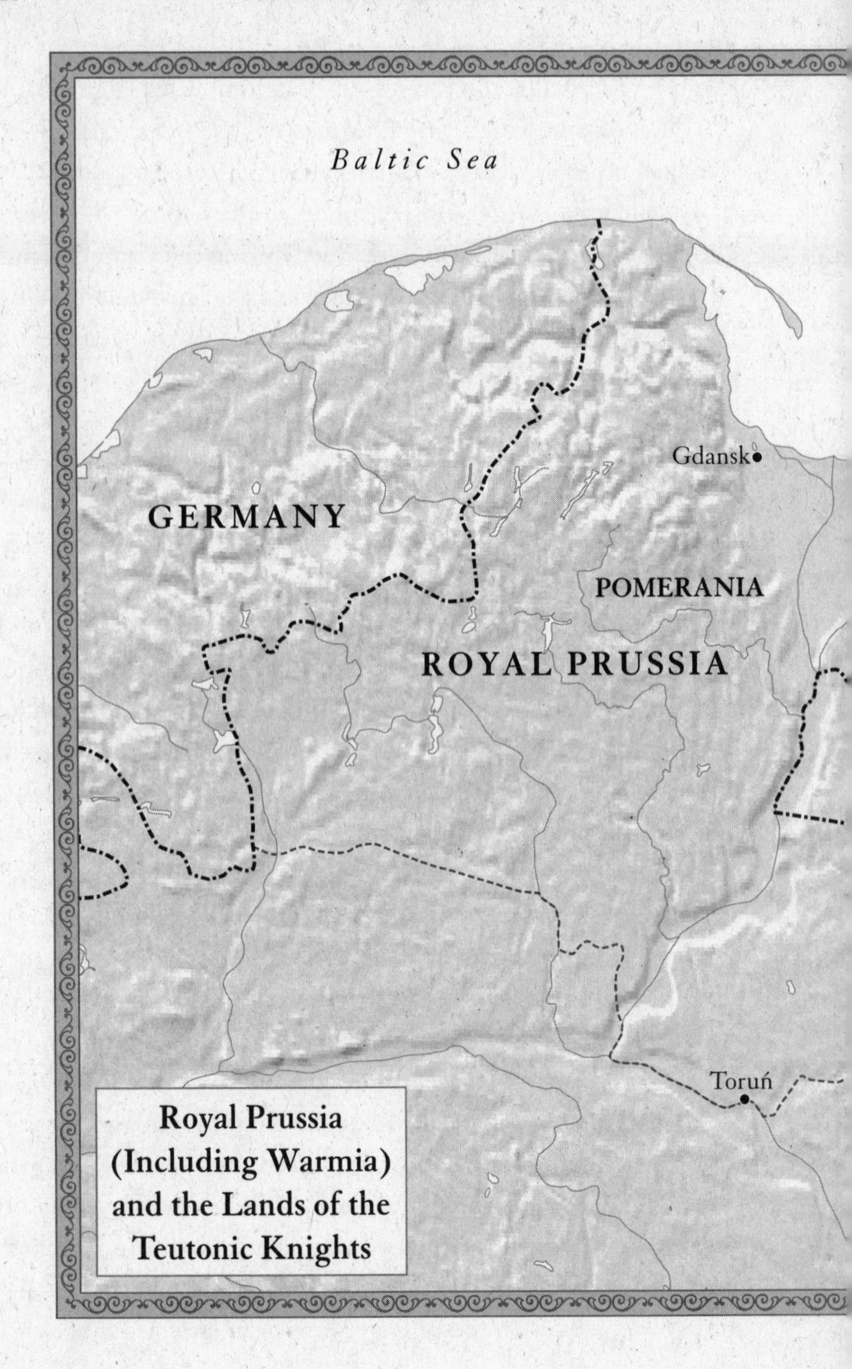

Royal Prussia (Including Warmia) and the Lands of the Teutonic Knights

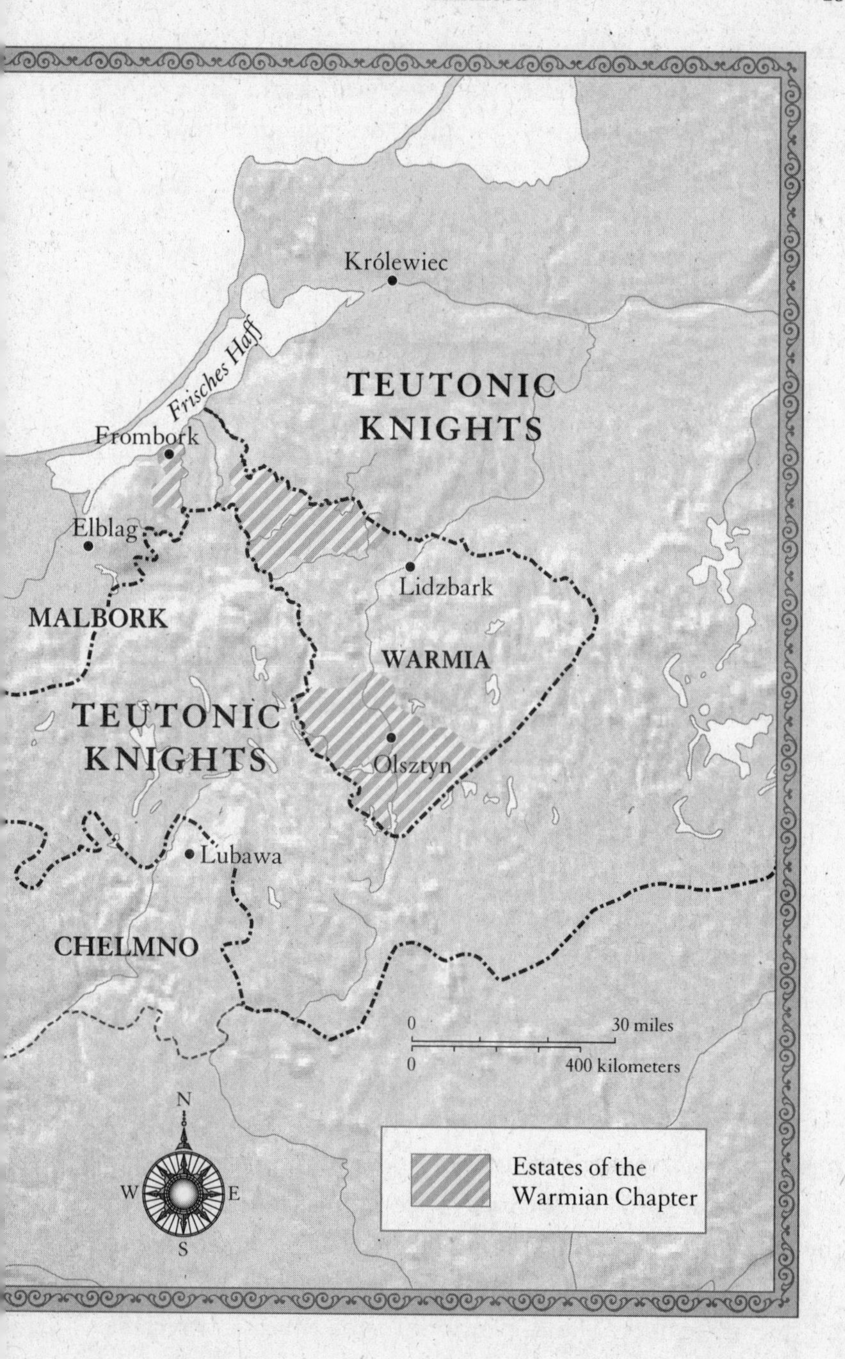
Królewiec
Frisches Haff
TEUTONIC KNIGHTS
Frombork
Elblag
Lidzbark
MALBORK
WARMIA
TEUTONIC KNIGHTS
Olsztyn
Lubawa
CHELMNO
0
30 miles
0
400 kilometers
N
W
E
S
Estates of the Warmian Chapter

the Roman Empire.) Amber was plentiful and the Knights controlled the trade completely, on penalty of death. (Ironically, perhaps the greatest demand for amber throughout Europe was for rosary beads.)

Though oppressive rulers, the Knights were impressive builders. They founded over seventy towns, and built large and sturdy defensive castles in each. Many of these castle fortresses still stand today, and they remain so completely intact that a knight returning from the grave would feel right at home. They were constructed of red brick, often with black bricks mixed in to create patterns. Usually designed in a square shape, they were invariably built on rivers and creeks to make defenses easier. Brick was the building material of choice because there is very little stone in the area, yet plenty of clay, the key ingredient for brick making.

The Knights, a monastic order of devout Christians, also built many churches, usually adjacent to a fortress. The castle-church combination would serve as a magnet for the farmers in the area and soon towns emerged filled with craftsmen. Within a few generations defensive walls were erected around the towns, and roads carved through the woods and fields to connect them. The cities and towns that the Knights founded in the thirteenth and fourteenth centuries remained important during Copernicus's life.

COPERNICUS'S ANCESTORS were native Germans who migrated east in the thirteenth century. They originally settled in the province of Silesia, in today's western Poland, probably in the town that is still called Koperniki. The family then moved on to Kraków, the capital of the Kingdom of Poland, in the mid-fourteenth century. The Kopernik surname appears in records there as early as 1350, and Copernicus's forebears may have included an administrator of the town's baths, an armorer, and a stonemason. The father of Astronomy's great-grandfather, also named Nicolaus, was made a citizen of Kraków in 1396. For several generations before Nicolaus's birth, the Koperniks were successful merchants of burger status, meaning that they were among the moneyed elite of Kraków.

Copernicus's father, yet another Nicolaus, left Kraków for the somewhat smaller city of Toruń in the late 1450s. Like his forefathers, Nicolaus was an astute and aggressive merchant. He must have smelled an opportunity in Toruń, because the city was in the midst of a war with the Teutonic Knights when he arrived. Toruń and the rest of the Prussian Estates (a clutch of cities and fiefdoms) were allied with the Kingdom of Poland against the Knights. By Nicolaus senior's time, the Knights were a bedeviling presence. Moreover, the kings of Poland had never recognized their sovereignty in the first place. The hostilities in what was later called the Thirteen Years' War started in Toruń in 1454, when its citizens rebelled against the oppressive rule of the Knights by storming their huge castle there, successfully taking it over, and killing or imprisoning the defenders.

Nicolaus senior thrived in Toruń. In every mention of his name in the official city records he is either receiving money owed him or else lending money, usually to the city itself. Most of his lending to Toruń occurred during the Thirteen Years' War, helping to pay and clothe the Polish soldiers and build a bridge over the Vistula. In addition to his successful business dealings, he also became a civic leader, serving as magistrate and alderman. He married very well—in the late 1450s or early 1460s he married Barbara Watzenrode. Barbara was the daughter of one of the wealthiest men in Toruń, Lucas Watzenrode. Nicolaus senior and Watzenrode had both been generous to the citizens of Toruń during the war, and this collaboration probably brought Nicolaus into the Watzenrode circle. When Lucas died in 1462, the Copernicus family inherited a great deal of money and property.

The Prussian Estates and the Kingdom of Poland finally defeated the Teutonic Knights in 1466. Because the Thirteen Years' War had started in Toruń, the treaty ending the conflict was also signed there. The Treaty of Toruń called for the incorporation of the Prussian Estates within the Kingdom of Poland. The Estates consisted of three independent cities (Toruń, Gdansk, and Elblag), and four bishoprics, which were small provinces ruled by prince-bishops. After the peace, this collection of territories became known as Royal Prussia. One of the bishoprics, called Warmia, became the province of Copernicus's maternal uncle Lucas Watzenrode in 1489. (The Bishop of Warmia was the first among equals within Royal Prussia, and he was the chairman of the Prussian Council, which included the leaders of the four territories, plus the mayors of the independent cities. Though Royal Prussia was part of Poland and the king was the sovereign, the king allowed Royal Prussia a great deal of independence.) Copernicus would spend his adult life in Warmia. The Teutonic Knights were allowed to keep one bishopric of their own (originally called Sambia), which nearly surrounded Warmia. Those ancient warriors were not quite finished causing trouble.

NICOLAUS COPERNICUS was the youngest of four children. He had an older brother, Andreas (believed to have been born in 1470), and two older sisters, Barbara and Catherine. Three of the four children would later find vocations in the Church—

Nicolaus and Andreas as canons, Barbara as a nun in the convent of Chelmno (she eventually became the prioress, or mother superior, there). Only Catherine married and raised a family.

The Copernicus family was extremely well off when baby Nicolaus joined them. Nicolaus senior was at least forty-five years old, and more likely closer to fifty. The family and its servants lived in a spacious, three-story townhouse, which it had inherited from Barbara's father, Lucas Watzenrode. The structure was built around the year 1350, and it was located near the center of town. It had an elaborately designed brick façade with large windows facing the narrow cobblestone street. There was a courtyard in the rear. The high ceilings, murals, and expensive furniture made it very comfortable. It was also well heated, with fireplaces throughout. Seven years later, Nicolaus senior, having grown in wealth and prestige, moved the family into an even larger, more impressive house. The new home was located right on the large town square. The family also owned a vineyard and other properties outside the city walls. There is no doubt that Copernicus and his brother and sisters led privileged childhoods.

Toruń was built alongside the Vistula River, which flows into the Baltic Sea about one hundred miles to the north. It was founded in the 1200s by the Teutonic Knights—Toruń was their first major settlement, in fact. The Knights built a huge defensive wall around the town, with thirty guard towers spaced evenly apart; in the middle of the wall, along the river, they built their first castle. By the late fifteenth century, Toruń was a thriving market town and trading hub with about 10,000 inhabitants, making it a large city for that era. It was now part of the Hanseatic League, a collection of cities that were part of an extensive trading empire. Huge regional markets were held each week in the town square, where Nicolaus and his siblings lived. The docks along the Vistula were busy with barges being loaded and unloaded as goods were moved up and down the great river. Wealthy landowners and merchants from across north central Poland transported their timber, coal, furs, wheat, and honey there for shipment to Gdansk; from that Baltic port the goods were sent all over northern Europe.

Toruń would have been a wonderful city to grow up in. As a prosperous trading center, all of the residents lived well. The three- and four-story town houses (some with plumbing) were built of brick and often adorned with decorative Gothic designs, just like the Copernicuses'. The clean, well-maintained streets were all paved with cobblestones. And each block had at least one church, inn, and granary. The first house the young Nicolaus lived in was only two blocks away from one of the town gates, beyond which flowed the waters of the Vistula. Though the Toruń side of the river was busy with the loading of barges, the other side, which was connected by the bridge that his father lent money to help build, would have been ideal for swimming and fishing. He and Andreas probably hunted for game with bows and arrows, too. A visitor later colorfully described this part of Europe:

> Apollo . . . is filled with a passion for hunting. For this reason he seems to have chosen this land before all others . . . For no matter where you turn your eyes, if you look at the woods, you will say that they are game preserves and beehives stocked by Apollo; if you look at the orchards and fields, rabbit warrens and birdhouses, lakes, ponds, and springs, you will say that they are the holy places of Diana and the fisheries of the gods. And Apollo appears to have chosen Prussia before other regions,

> I say, as his paradise. Besides stag, doe, bear, boar, and the kind of wild beast that is commonly known elsewhere, he brought in also urus, elk, bison, etc., species scarcely to be found in other places, to say nothing of the numerous and quite rare types of bird and fish."

Within the city walls, the huge town square became Copernicus's front yard when the family moved into their second home. The town square was a jewel—a huge expanse of cobblestone surrounded by lovely buildings, with the massive and ornate town hall sitting right in the middle. The square teemed with life—the men wearing their colorful fur-trimmed robes over tunics, britches, and boots, and the women adorned in their long, layered dresses. And just a few blocks away from the square was the recent ruin of the huge castle of the Teutonic Knights that had dominated the city right up until the rebellion of 1454. The children of Toruń no doubt used the ruin as a playground (as they still do).

A fifteenth-century chronicler said this about Copernicus's town: "Toruń with its beautiful buildings and its roofs of gleaming tiles is so magnificent that almost no other town can match it for beauty of location and splendor of appearance."

WHEN NICOLAUS WAS ABOUT TEN, he and his siblings lost their father. Fortunately, their mother's brother, Lucas Watzenrode (1447–1512), was already wealthy and becoming established in the hierarchy of the Church. In 1483, he would be made a canon of the Frombork Chapter, and in 1489, he would be elected bishop of Warmia. He became the benefactor of his sister's children (at the time he had no children of his own). As a lad, Nicolaus attended grammar school and "higher" school, perhaps at nearby Saint Johann, both of which were under the supervision of the Church.

At this stage, there was no indication that young Nicolaus was destined for fame. His father and grandfather had been successful businessmen, which meant that they were astute thinkers. But that is all we know. In his few writings, Copernicus gave no hint as to what motivated him to study the night sky so intently. As mentioned in the previous chapter, Regiomontanus was constructing astronomical tables as a teenager; there is no evidence that Copernicus displayed a similar predisposition.

One intellectual pursuit that may have sparked his interest in the heavens was the then universal belief in astrology. The night sky was vivid to observers five hundred years ago, before the electrified world of our modern age muted it with ambient light. On clear nights, with only a lantern or torch to provide just enough light to walk without stumbling, one had the stars and moon as constant companions, so brilliant that they seemed to be almost within reach. Aristocrats, gentry, and peasants alike were surely curious about the black curtain that blanketed their world each night, just as they were inquisitive about what the sun revealed during the daylight hours. And there were individuals who no doubt looked at the dark canopy and pondered why the specks of light moved the way they did. But for the vast majority of those who strove to understand the heavens—why its components were there in the first place and what their movements signified—the motivation was not just the accumulation of basic knowledge; rather, it was to find meaning in these motions and patterns.

In an era in which God was believed to be an active participant in the lives of his creations, when the earth was assumed to be the middle of the universe, and when it was believed that all of the objects in the night sky were there only for God's creatures to gaze at, people assumed that the heavens contained communications. Why else were these objects there? The only one that had any tangible impact on the earth was the moon, which gave off light and affected the tides. The stars and wandering stars were there to be read. Even the Bible proclaimed it: "And God said, Let there be lights in the firmament of the heaven to divide the day from the night; *and let them be for signs*, and for seasons, and for days, and years" (emphasis added).

It is difficult for citizens of the twenty-first century to understand how important the field of astrology was for citizens of the fifteenth and sixteenth centuries. Today, astrology is a quaint and whimsical anachronism. But in Copernicus's era, astrological predictions and horoscopes were sought and embraced almost as gospel truth. And when predictions later proved to be wrong, which was most of the time, the blame was ascribed to the obscure movements of the planets and stars, not the practice of astrology itself.

Whether young Nicolaus Copernicus was motivated by astrology is mere speculation, but within a few years he would be immersed in the study of the heavens. It is entirely possible that his fascination started in Toruń. In the mind's eye, it is easy to picture the teenage Nicolaus spending the night with his brother at the family vineyard outside of Toruń. As the sun set and darkness started to settle over the gentle Polish countryside, Nicolaus would notice the first stars appearing on the horizon. He knew that they were called the wandering stars because of their constantly shifting positions. Of course the moon rose in a slightly different spot and time, crossed the sky along a different path, and was always a bit more or less illuminated than the night before. As the night wore on, if he and Andreas were paying enough attention, they would have noticed the constellations moving, too—in this case they held their relative positions as they pivoted around the North Star. What did this dance of the night sky mean?

4

Student Years

OVER TIME, Uncle Lucas resolved that the Copernicus brothers should pursue university educations and start to see the world. Lucas had been a student at the University of Kraków in the 1460s, and he decided that this was the appropriate institution for his nephews. In the fall of 1491, Andreas and Nicolaus left their home, presumably for the first time. Kraków, also on the banks of the Vistula River, was far to the south of Toruń. The brothers probably traveled upriver by boat. They likely spent the better part of a week aboard a "lodi," which was a barge equipped with a wide sail—it was an all-purpose boat used throughout the Hanseatic League towns to move goods and people.

Kraków was beautiful. Its town square dwarfed Toruń's, and the wide streets revealed elegant churches and town houses topped with fancy cupolas around every corner. Wawel Castle, where the king reigned, stood like a sentry on a hill overlooking the town. Kraków and its citizens were rich from trade; the city was located at the intersection of the Prague-Crimea road and the amber route between Gdansk and southern Europe. It had twice as many people as Toruń. The diverse population included native Poles, Germans, Lithuanians, Italians, Hungarians, and Jews. The Jewish population was one of the largest in Europe, but at about this time the Jews were being pushed out of the city to a neighborhood on the outskirts.

Central Europe at the Time of Copernicus

Nicolaus was almost nineteen when he enrolled at the University of Kraków, considered old to be starting university studies by the standards of the fifteenth century. Andreas was even older. The university still has the document from late 1491 on which "Nicolaus Nicolai de Thuronia" and "Andreas Nicolai de Thorun" officially registered as students; young Nicolaus was the thirty-second name on a list of sixty-nine incoming students. Until Andreas's death in 1518, the two brothers from Toruń did everything together. Historians have had difficulty acquiring details about Andreas Copernicus, but it appears that he was the fun-loving, less-talented, older brother of the more serious, gifted Nicolaus. Though Uncle Lucas treated them equally, Nicolaus was consistently taken care of first. The first hint that they had different priorities comes from that same student register. Nicolaus paid his tuition in full and Andreas did not—presumably Uncle Lucas provided enough funds for the full payment, but Andreas had other plans for the money.

The University of Kraków had been founded by the king of Poland, Casimir the Great, in 1364, making it the second oldest in Central Europe—behind only the University of Prague, founded in 1348. (The University of Paris and several Italian universities—Bologna, Padua, and Naples—were older.) The university was in the center of Kraków itself. It was housed in a large square building of several stories with a spacious courtyard in the middle. Most of the students were sons of rich merchants and landowners; sons of peasants could matriculate to a university only if the landowner they labored under paid for them, so they were very rare. About half the students during Copernicus's time at the university were from outside Poland, predominantly from Hungary, the German principalities, Scandinavia, and even Italy.

Most of the 800 or so students took rooms at the seven

student hostels, or dormitories (called *bursae*), where they lived together near the university. Yet, it appears that the Copernicus brothers took rooms at the home of Piotr Wapowski, who was a friend and fellow canon of Uncle Lucas. While under this roof, Copernicus formed a lifelong friendship with Piotr's son, Bernard Wapowski, who would go on to become a renowned cartographer.

The main curriculum at the University of Kraków was similar to that at the other European universities. Thus, nearly every student took courses in law, Latin, rhetoric, philosophy, and theology. However, from nearly its inception the University of Kraków placed an emphasis on astrology and astronomy, and therefore had an unusually large number of mathematics and science courses. The university had a chair in astrology and astronomy as early as 1410, the first such chair in central Europe. In fact, the Nuremberg geographer Hartmann Schedel, in a popular chronicle published in 1493, noted the fame of the University of Kraków as a center for the study of astrology and astronomy: "There is in Kraków a famous university, which boasts many most eminent highly educated scholars and in which numerous liberal arts are practiced. But the science of astronomy stands highest here, and in all Germany there is no more renowned university in this respect, as I know exactly from the narrations of many persons."

From the classes that he took and the books that he bought, it is clear that Nicolaus Copernicus immersed himself in astrology and astronomy during his University of Kraków years. Records indicate that he may have taken up to seven separate courses in these fields, a number that would make even a modern astronomy major envious. The courses had titles like "The Spheres," "Euclid's Geometry," "Planetary Theory," "Tables of Eclipses," and, of course, "Astrology."

When Copernicus matriculated, the chair in astronomy, and thus the chief astronomy scholar at the university, was Adalbert of Brudzewo (1445–1497), who may have studied under Regiomontanus at the University of Vienna. Adalbert was no longer teaching astronomy (he was teaching philosophy) when Copernicus was a student, but his published books and course notes were used in many classes, and he still held Saturday astrology/astronomy workshops at his home. Copernicus was also able to take courses in meteorology, surveying, and geography. His geography professor, Matthew of Miechow, was one of the most accomplished geographers in Europe and the author of a standard text.

The early 1490s were remarkably eventful years in Kraków and at the university. The old king, Casimir IV, passed away in 1492 and the spectacular coronation of his successor took place a few months later. That same year there was a fire at the university that destroyed part of the main building. As the capital of the then-flourishing Kingdom of Poland, Kraków attracted scholars, traders, entertainers, and wealthy visitors from all over Europe—in 1494, a sultan visited the city with sixteen camels in tow.

Something else happened while Copernicus attended the University of Kraków that affected everyone in Europe, not just the citizens and undergraduates in the Polish capital. Christopher Columbus landed in the Bahamas and then Cuba in October of 1492. He returned and docked at Lisbon in March 1493—just after Copernicus turned twenty years old—and the news he brought back spread across all of Europe within weeks. At first, it was not entirely clear what Columbus had discovered, but the scholarly community knew that it was important. From this point on, the world was a different place. Columbus's voyage proved once and for all that the earth was not flat. More stunning, though, was the realization that there was another part of the world that had previously been unknown, and that mysterious place contained strange natives. Upon Columbus's return there was an explosion of interest, and demand for experts, in cartography (both maps and globes), surveying, navigation, and also astronomy.

For one interested in astrology and astronomy, every day was a new adventure. Martin Bylica, an astronomy professor at the University of Buda and the one who had drawn Regiomontanus to Buda in the late 1460s, had been a student at Kraków and he sent several state-of-the-art astronomical instruments to his alma mater in 1492. The students started using them immediately. Even the heavens themselves cooperated—during the four years that Copernicus lived in Kraków, there was an unusual number of astronomical phenomena to view. A comet was visible in central Europe in late 1492 (another one had appeared in January 1491, before the Copernicus brothers enrolled at the university). In October 1493, there was a partial solar eclipse; and in 1494, there was another solar eclipse and two lunar eclipses, one occurring on Good Friday. Thus, Copernicus had the good fortune to be present at an unusual confluence of events—there was much happening in the world (in addition to Columbus's discovery and the succession in the Polish monarchy, the pope died in 1492), in Kraków and at the university, and in the night sky. How energizing for a serious and bright young man like Copernicus.

The future father of astronomy bought his first books during this stimulating period of his life. The volumes were obviously chosen carefully and considered prized possessions. They followed him everywhere he went for the next five decades of his life, and were still in his personal library when he died. Books were very expensive at this time and represented a significant purchase for a constantly cash-starved student. He bought copies of Euclid's *Elements* (1482), the *Alfonsine Tables* (1492), and Regiomontanus's *Tables of Directions* (1490). Copernicus had the latter two bound together along with sixteen blank pages in the back; several of the blanks contained tables from Peurbach's *Table of Eclipses*. The *Alfonsine Tables* were the standard source for charting the wandering stars against the fixed stars in the night sky. Recall that the Regiomontanus volume contained projections for the daily rotation of the heavens. From the notes and observations he recorded on the blank pages in the years to come, it is evident that Copernicus used his book of tables extensively. Nicolaus's purchase of these expensive books while still an undergraduate reveals how serious he was about astrology and astronomy at this early stage.

In 1495, probably at the end of the spring semester, Nicolaus and Andreas left the university after nearly four years of study. Many scholars assume that they each received their baccalaureate degree, but there is no such record for either one. The brothers returned to Warmia and probably lived with their uncle Lucas, who was now the bishop of Warmia. As the ruler of that bishopric, he lived in a magnificent castle at Lidzbark. One year later, by the late summer of 1496, Nicolaus was on the move again. For the first time Copernicus left his homeland of Poland. (It is not known if Andreas accompanied him.) On foot and horseback, he traveled the roads and slept in the inns of central Europe for about two months. Then he took up residence in the intellectual center of the world at the absolute peak of its ascendancy—Italy during the time of Leonardo and Michaelangelo.

In October 1496, Nicolaus enrolled at the University of Bologna, one of the oldest and most prestigious universities in Europe, intending to acquire a doctorate in canon law. Andreas followed Nicolaus to the University of Bologna two years later, enrolling in the fall of 1498, also to acquire a doctorate in canon law. A generation before, Uncle Lucas had done precisely the same thing, obtaining his doctorate in canon law in 1473. Thus young Copernicus was following precisely in his uncle's footsteps. From 1496 to 1503, Copernicus lived and studied in Italy. The broad outline of this eight-year sojourn is known, but many of the specifics are not. One thing is certain, though—this was the second important stage in his development as an astronomer.

His timing could not have been better, for earlier in 1496 the astronomy community celebrated the publication by a Venetian publisher of the *Epitome of the Almagest* by Regiomontanus. As mentioned earlier, this book had been long awaited. In fact, the publisher who finally succeeded in bringing it out had argued to the Venetian authorities who controlled the granting of copyrights that "this work is rare and also has been seen by very few scholars. This is because everybody who can get hold of it keeps it hidden as his own treasure to prevent other specialists from asking to borrow it."

Copernicus must have borrowed and read this beautiful volume with profound interest. In the *Epitome* Regiomontanus boldly stated that Ptolemy's theory of the motion of the moon could not be correct. If it were, then the moon should vary much more in its appearance. Ptolemy stated that the moon's orbit around the earth regularly took it twice as far from the earth as at other times. This clearly does not happen. For an active astronomer like Copernicus, perhaps already skeptical of the Ptolemaic model based on what he had learned in Kraków, Regiomontanus's criticism was pivotal.

Another important work appeared at about the same time as the *Epitome*. This was titled *Disputations against Divine Astrology* (1496), written by Giovanni Pico della Mirandola (1463–1494). The book was an aggressive attack on astrology. One part of the criticism that Copernicus certainly paid attention to was Pico's assertion that if astronomers argued about the order of the planets—the relative positions of Mercury and Venus were in dispute—confirmation of which was a fundamental necessity for astrological prediction, how could the predictions be worth anything? Copernicus was notably silent about astrology through most of his later career, and without a doubt was focused on pure theoretical astronomy in the years ahead. Perhaps his lack of interest in astrology started with his reading of Pico.

While taking the required classes in canon law at the University of Bologna, Copernicus continued to pursue his study of the heavens. He boldly went right to the top and rented rooms from the leading astronomer-astrologer in Bologna, who happened to be one of the most accomplished in Italy. Domenico Maria de Novara (1454–1504), who like Adalbert of Brodzewo may have been a former student of Regiomontanus, had been professor of astronomy at the University of Bologna for more than twenty years. He was required to produce and publish an astrological almanac each year, which contained projections of the phases of the moon, positions of the planets, the date of Easter, and then his predictions of "good versus evil" days, depending on the heavenly readings.

In his almanac of 1489 Novara made an iconoclastic claim that was later repeated by many astronomers—that the earth's axis of rotation had shifted, and that it followed a slow movement that required many years to complete. To predict any motion of the earth's axis was significant, and any movement of a supposedly *perfectly still* Earth even more so. Novara was an original and skeptical inquirer.

Copernicus later said that he "was not so much the pupil as the assistant and witness of the observations of the learned Dominicus Maria" de Novara. It was also reported later that Copernicus "lived with Messer Maria of Bologna, whose calculations he knew exactly and at whose observations he assisted."

Copernicus the astronomer made his official entrance on the celestial stage on March 9, 1497, when he made his first known observation with Novara. At 11:00 P.M. they witnessed the eclipse of a bright star, Aldebaran, by the moon. This type of eclipse is called an occultation. The observation was used by Novara in his astrological forecast for the next year, and it was later used by Copernicus in *On the Revolutions* to provide proof for his theory of the motion of the moon.

Copernicus soon began making his own observations. In early 1500, he recorded studying two conjunctions of the moon with Saturn (they occurred on January 9 at 2:00 A.M. and March 4 at 1:00 A.M.). Copernicus made notes of his observations on the blank pages of his *Alfonsine Tables* volume. He spent some months in Rome in the second half of 1500. He later said that while in Rome "he lectured on mathematics before a large audience of students and a throng of great men and experts in this branch of knowledge." While there, he also made his fourth official observation, a lunar eclipse on November 6, 1500, at 2:00 A.M..

Copernicus also received his first serious instruction in Greek while at Bologna. As Regiomontanus had recognized before him, knowledge of Greek was essential for the serious astronomer to study the original sources. Copernicus bought a copy of a Greek dictionary at this time and retained it for the rest of his life.

* * *

IT WAS DURING the Italian period that Uncle Lucas gave Nicolaus lasting security—he arranged for his election (in absentia) as canon of the Cathedral Chapter of Warmia in 1497. Thus he became one of sixteen canons officially based in the small village of Frombork on the Baltic coast. Copernicus received official confirmation of his election on October 20, 1497 (Andreas received a similar appointment two years later). In early 1501, Nicolaus and Andreas traveled across the European continent and reported to the chapter in Frombork for the first time, even though they had been drawing income as canons for several years. While facing their future colleagues they asked for permission to continue their studies, Nicolaus in medicine, Andreas in law. Permission was granted. Nicolaus journeyed back to Italy and attended the University of Padua for the next two years, where he studied medicine.

The University of Padua was the finest medical institution in Europe. It was also Europe's oldest university, founded in the twelfth century. Medicine at that time was nothing like the modern practice. For instance, in the early sixteenth century a medical student did not even take an anatomy course. However, Copernicus *was* exposed to medical astrology, which meant paying attention to horoscopes and what they conveyed about a patient's overall constitution, as well as what the stars and moon said about general conditions (was it a good day or bad day for a treatment?) and the likelihood of an epidemic. Some Copernicus scholars believe that he learned something about the Arab astronomical tradition while in Padua. Arab astronomers exhibited more skepticism about Ptolemy than European astronomers.

Copernicus finally finished his higher education in 1503,

when he received a doctorate in canon law from the University of Ferrara in May of that year. He spent very little time at Ferrara, and Copernicus scholars have speculated that he chose to finish there because it was a cheaper institution from which to graduate. A graduating student had to host and pay for a huge celebration upon completing his degree.

The attainment of his doctorate meant that Copernicus's traveling student days were now behind him. This had been a remarkably stimulating and fruitful period of his life, and he must have been hesitant to end his studies. However, even greater intellectual adventures awaited him.

5

WARMIA

DEGREE IN HAND, the thirty-year-old Nicolaus Copernicus left Italy and returned to Poland permanently in the summer of 1503. He had spent the previous twelve years in Kraków, Bologna, Rome, Padua, and Ferrara, right in the middle of the intellectual ferment of the Renaissance. From this point on he would live about as far from the major centers of European learning as possible. He had interacted with one world-class astronomer, Domenico Maria de Novara, and many other enlightened scholars at the universities he had attended. Now he would work more or less in isolation. (As mentioned in the first chapter, it is curious that Copernicus was not offered a post in Italy as a university professor or court astronomer/astrologer. Perhaps his talents in mathematics and astronomy were not well known—his actions later in life indicate that he was guarded, at least in conversation, about his research. Or, perhaps he really was a late bloomer. His first observation came only in 1497. We have no information about how the relationship between Copernicus and Novara ended; all that is known is that Nicolaus moved on to another university.)

Copernicus's first post after returning was as his uncle Lucas's personal secretary. He performed these duties for seven years, residing with the bishop at his palace in the small town of Lidzbark. In 1510, he left his uncle's side and finally settled in Frombork, becoming part of the chapter of canons. Frombork would remain his home for the rest of his life, with the exception of one five-year absence. Though he traveled extensively in Warmia and Prussia, his small corner of the world, Copernicus's adult years were settled in comparison to his youth. It was the first years of this period of his life that were his most creative.

SETTLED LIFE agreed with Nicolaus Copernicus, at least as far as his astronomical studies were concerned. Over the course of the next ten years, from 1503 to about 1513, he developed the heliocentric theory of the heavens. We know this from a report, or memorandum, that Copernicus drafted and circulated among his astronomy-minded friends sometime before 1514. Matthew of Miechow (1457–1523), one of Copernicus's professors at the University of Kraków (he taught geography), made a careful inventory of his personal library dated May 1, 1514. Part of that inventory included a "manuscript of six leaves expounding the theory of an author who asserts that the earth moves while the sun stands still"

The report is now called the *Commentariolus*, but it is not certain what title, if any, Copernicus gave it, or whether he even identified himself as the author. Most likely it carried no title and was anonymous. Written in Latin, it was drafted on a clutch of large sheets of paper. To read it today is an amazing experience, especially knowing the astronomical beliefs of the time. For the boldness with which the canon from Frombork pronounced his new model of the universe is astounding. He confronted nearly every accepted fact about the heavenly bodies, facts that had been accepted since Aristotle's day and formalized by Ptolemy, then inculcated through the teachings of the Church.

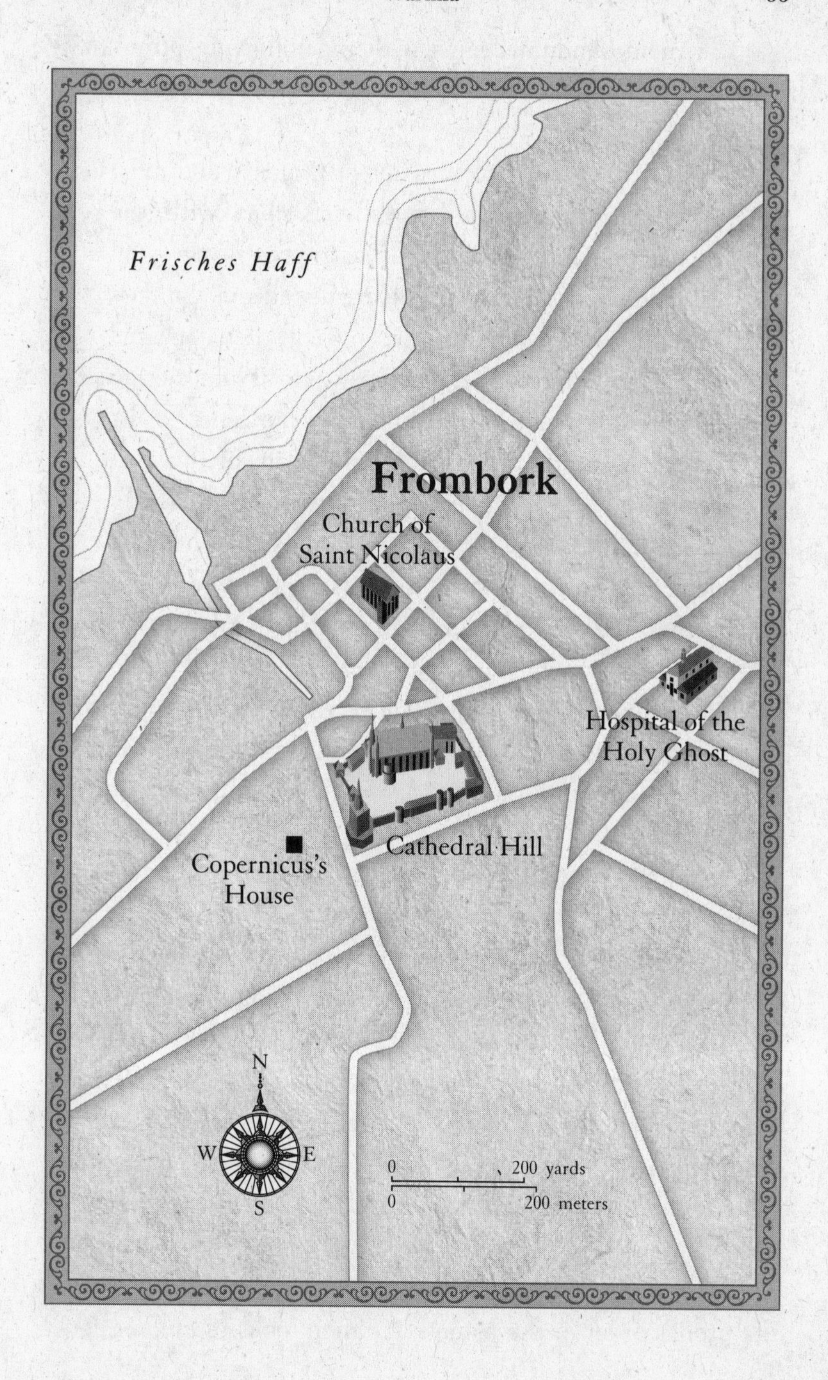
Frisches Haff
Frombork
Church of
Saint Nicolaus
Hospital of the
Holy Ghost
Cathedral Hill
Copernicus's
House
N
W
E
S
0
200 yards
0
200 meters

The memorandum begins with the following pronouncement:

> Yet the planetary theories of Ptolemy and most other astronomers, although consistent with the numerical data, seemed likewise to present no small difficulty. For these theories were not adequate unless certain equants were also conceived; it then appeared that a planet moved with uniform velocity neither on its deferent (main orbit) nor about the center of its epicycle (second orbit). Hence a system of this sort seemed neither sufficiently absolute nor sufficiently pleasing to the mind. Having become aware of these defects, I often considered whether there could perhaps be found a more reasonable arrangement of circles, from which every apparent inequality would be derived and in which everything would move uniformly about its proper center, as the rule of absolute motion requires.

Then, barely two pages into the essay, Copernicus asks the reader to grant him seven assumptions, which he calls axioms. The second axiom is, "The center of the earth is not the center of the universe." The third axiom is, "All the spheres revolve around the sun as their midpoint, and therefore the sun is the center of the universe." The fifth axiom is, "The earth . . . performs a complete rotation on its fixed poles in a daily motion, while the firmament and the highest heaven abide unchanged." One cannot stress enough how revolutionary these three pronouncements were.

After listing all of his axioms, Copernicus goes on to state the proper order of the planets. He then looks in greater detail at the motions of the earth, the moon, the "superior planets" (Mars, Jupiter, and Saturn), and finally Venus and Mercury.

Here is a tally of what Copernicus's new heliocentric theory propounded:

1. The earth and the other planets revolve around the sun, not the earth.
2. The moon is the only heavenly body that *does* revolve around the earth, thus separating the moon from the other wandering stars.
3. The earth rotates on its own axis once every twenty-four hours.
4. The revolutions of the outer planets take much longer than Ptolemy thought—Saturn takes 30 years to revolve around the sun, Jupiter 12 years, Mars 2 years, Earth 1 year, Venus 9 months, and Mercury 3 months (Copernicus's estimates were essentially correct and he placed the planets in their proper order for the first time.).
5. The universe is profoundly larger than previously believed, with the "firmament," or fixed (nonwandering) stars, so far away as to make the "distance from the earth to the sun . . . imperceptible in comparison with the height of the firmament."

Though the *Commentariolis* was quite short, it was clearly a work of mature thought and years of observation, as well as the study of older books. There were no mathematical proofs in it, but the precision and detail found in the document would lead a reader to believe that the author indeed had proof of his assertions. Early in the piece Copernicus states, "However, I have thought it well, for the sake of brevity, to omit from this sketch mathematical demonstrations, reserving those for my larger work." The larger work is a reference to the manuscript that would become *On the Revolutions of the Heavenly Spheres*. Thus it appears that Copernicus started his great book at least thirty years before it was published.

How did Copernicus arrive at the revolutionary heliocentric concept? The only direct statement that he made, both in the *Commentariolus* and later in *On the Revolutions*, is that Ptolemy's use of the equant was problematic because his model did not exhibit "uniform circular motion." This serious flaw in Ptolemy's conception had bothered the Arabic astronomers in the centuries before Copernicus, too. As a rigorous scientific thinker, Copernicus could not accept the fact that the cosmos could behave without obeying first principles—specifically uniform circular motion. When Copernicus placed the sun in the middle of the planetary orbits, as he would later say, the pieces all fit together. "Not only do all their phenomena follow from that but also this correlation binds together so closely the order and magnitudes of all the planets and of their spheres or orbital circles and the heavens themselves that nothing can be shifted around in any part of them without disrupting the remaining parts and the universe as a whole."

The tone found in the *Commentariolus* shows that, at least when it came to his astronomical studies, Copernicus was supremely confident. In fact, he sounded like Regiomontanus. Early, he comments, "After I had addressed myself to this very difficult and almost insoluble problem, the suggestion at length came to me . . ." And there are passages like "my explanation is the truer" and "those who employ . . . fall into two manifest errors" throughout. Finally, toward the end of the report he states, "Yet Mercury too will be understood, if the problem is attacked with more than ordinary ability."

* * *

PROBABLY BECAUSE OF the *Commentariolus*, in 1514, Copernicus was one of the astronomers officially invited by a committee appointed by Pope Leo X to participate in the reform of the calendar. The Julian calendar had been in effect since the reign of Julius Caesar, when he took the bold step of radically reforming the Roman calendar, which had become out of sync with the seasons. Caesar made a year 365 days long, with a leap year every fourth year. Though the Julian calendar was similar to our modern one, small discrepancies between the calendar and the annual revolution of the earth around the sun—only about eleven minutes a year—had added up (the eleven-minute annual discrepancy equaled five days in a millennium), and the actual weather of the seasons was clearly off (winter came earlier than it should, as did spring). By the early sixteenth century there was pressure to make a major adjustment.

Recall that Regiomontanus had been called to Rome for the reform of the calendar in 1475. The Nuremberg astronomer's untimely death had brought the earlier reform effort to a halt, and it was only revisited in the second decade of the sixteenth century. To be invited by the committee for this important task indicates that Copernicus was now known in astronomical circles as an outstanding member of their community. Copernicus did send a response to the committee, but it has not survived and it is not known what he wrote. In the end, the calendar was not reformed this time, either.

Between the circulation of the *Commentariolus* and the publication of *On the Revolutions* many years later, Copernicus wrote only one other astronomical document that was seen by outside eyes. Now called "The Letter Against Werner" and drafted in 1524, it was a long formal letter that essentially reviewed a book by the Nuremburg astronomer and mathematician Johannes Werner (Copernicus had been asked to review it by another scholar, Bernard Wapowski, a friend since their University of Kraków days). Copernicus was deeply critical of Werner's ideas about the Eighth Sphere, which was the term given to the fixed stars. Toward the end of the letter he again refers to his larger work: "What finally is my own opinion concerning the motion of the sphere of the fixed stars? Since I intend to set forth my views elsewhere, I have thought it unnecessary and improper to extend this communication further."

THERE ARE TWO REASONS why the immensely gifted Copernicus did not produce a greater number of written works in his early adult years. The first is that *On the Revolutions* was an extraordinarily complex project and it required nearly all of his scientific concentration. After all, he was attempting a work to rival a book—Ptolemy's *Almagest*—that had not been superseded in more than 1,300 years. The second is that unlike the great astronomers who preceded him—Peurbach, Regiomontanus, Novara, and others—Nicolaus Copernicus was not an academic with the freedom to devote all of his energies to his studies. He had a day job, as it were, or more properly, a series of day jobs, that took a great deal of his time. All of Copernicus's future vocations stemmed from his appointment as a canon in the Church. Though he lived comfortably the rest of his adult life, his canonship would eventually cause him a great deal of strife and would nearly derail his life's work.

As described earlier, a canon was a member of the clergy of the Church—he was required to take "first orders" and the vow of celibacy. To become a priest, which was encouraged but not mandatory, a canon had to take higher orders. Copernicus was one of sixteen canons attached to the cathedral in Frombork in the bishopric of Warmia. Recall that his uncle Lucas, the bishop, secured the canonship for Nicolaus in 1497 while he was a student in Italy. Following Poland's defeat of the Teutonic Knights in 1466, the king allowed the territories collectively known as Royal Prussia or the Prussian Estates to continue to govern themselves with the systems already in place. This meant that bishops ruled Warmia, Chelmno, Marlbork, and Pomerania. The bishop of Warmia was the first among equals. He was the chairman of the Prussian Council, which included the leaders of the four territories, plus the mayors of the independent cities of Gdansk, Toruń, and Elblag.

Warmia was a wedge—it was actually shaped like a triangle—that consisted of approximately 2,500 square miles of flat, arable land. Like all of northern Europe, the region was a low-lying plain that had been covered by the great ice sheet of the last ice age. The retreating glaciers left a scraped landscape characterized by numerous small lakes. The area was also heavily forested. The tip of the wedge was the small town of Frombork on the coast of the Baltic Sea, where the canons resided.

As a group, the canons were called the "Chapter of Warmia." Most of the sixteen lived in the coastal town, though two or three were absentee, as they set up residences elsewhere either in Warmia or in Rome, where they represented the chapter's interests. They led a half-religious, half-secular existence. The religious part of their lives took place within the high red brick walls of the cathedral complex atop the small hill overlooking Frombork. It was called, appropriately enough, Cathedral Hill. Although only sixty-five feet above sea level, Cathedral Hill dominated the surrounding flat countryside, and the 30- to 50-foot-high ramparts could be seen from quite a distance. Rising above the fortress walls was the spectacular Gothic cathedral, formally called the Cathedral of the Ascension of Our Lady Mary and St. Andrew Apostle. Completed in 1388 and built with red brick and tile, it was 300 feet long, 75 feet wide, and 50 feet high, and it rivaled the finest cathedrals in Europe. Natural light poured through the numerous windows, illuminating the gold and marble that filled the nave. Each canon had his own small altar, situated between the pillars that supported the massive structure.

Adjacent to the cathedral and built into the defensive walls was the bishop's mansion, which he used when he visited Frombork. On the other side of the cathedral was the dormitory in which most of the canons had an apartment. Copernicus had a different arrangement, though. His apartment was located in one of the tall protective towers. He actually purchased it from the chapter and occupied all three stories of the structure. There was a catwalk along the rampart that provided an impressive view of the small town below and the sea beyond.

In addition to their rooms within the cathedral fortress, each canon had at least one house outside the cathedral walls, called a *curia,* and most also had a small villa in the countryside, called an *allodium* or grange. The *curias* were spacious houses—not quite big enough to be called mansions. They were nonetheless impressive domiciles, and roomy enough for the canons and their servants.

All of the canons' *curias* were on Cathedral Hill, just outside the cathedral walls. So the canons lived with one another when inside the fortress, and they were one another's neighbors when outside it. Copernicus bought his house in 1514. His was in a group of three that shared a joint lawn probably half an acre in size. Copernicus's was on the edge of Cathedral Hill, overlooking Frombork, just a few dozen yards from the west gate of the cathedral fortress. Today a pedestrian has to walk across a ravine to get to the gate, but it is thought that a footbridge joined it to Copernicus's yard then.

The canons ran their own self-contained community. Within the walls or just outside them, they controlled a mill, bakery, malt house, brewery, and brickyard. The workers who labored in them were residents of the village of Frombork, which was dwarfed by the buildings on the Cathedral Hill that rose above it. It was a complete mismatch—the cathedral fortress should have been in a much larger city, not casting its shadow on a town of less than a thousand inhabitants. The citizens had their own church, St. Nicolaus. The presence of the Church in the lives of the townspeople of Frombork was palpable. In the evening, when they were doing their final chores before retiring for the night, they would look up at the hill and see the massive structure silhouetted against the night sky. Depending on how the canons were treating them at the time, it was either a wonder to behold, or an unwelcome reminder of the canons' control over their lives.

The townspeople spoke mainly vernacular German. They lived and worked in the houses and shops that lined the narrow streets between the base of Cathedral Hill and the docks on the Baltic Sea. Actually, it was not quite the Baltic Sea—the currents of the Baltic had created a barrier island that extended for about fifty miles. The body of water between the island and the mainland was called the Frisches Haff. (Today it is called Vistula Bay. Geologists believe that the sandbar formed approximately 5,000 years ago.) Frombork was on the edge of the mainland, separated from the sea by the Haff. The major industry in Frombork was fishing, and its docks were bustling with activity all day long. A visitor in the 1530s described the scene:

> In the summer more than 500 Zesekhan [net boats] are made ready. These are small trawlers which can sail against the wind as well as with it. The fishermen let out a net, called by them Zese, and with this spread out behind them they sail up and down the Haff. The larger fish are at once salted or else brought fresh to market in the cities along the Haff. But the small ones are dumped overboard, and it is a pity to see so many of them go to waste.

Of the general area, the same observer noted:

> "It is a level, plain land without any hills . . . full of rivers which flow in the direction of the ocean and are all navigable. The soil holds no ore deposits . . . nor are there vineyards, except around Stettin. Grapes could be grown here, but the inhabitants are so shiftless that they refuse to go to the trouble of cultivating them. Beers seems to be all they want. Other than that, the land grows a great plenty of grain and fruits, and there are forests and pastures where cattle are raised. Thus the people lack nothing. . . . For this reason the country is well populated and has thriving cities.

WHEN IN FROMBORK, the canons spent at least some time within the fortress walls each day. At a minimum, they were expected to attend and help serve mass twice a day, in the early morning and early evening. At mass, they wore vestments—there is a record of young Copernicus, who had not yet taken up residence at Frombork, being delinquent with the eight marks needed to pay for his vestments.

The church-related duties of a canon were neither onerous nor time-consuming. The secular duties outweighed them in terms of time and priority. Thus, though technically a clergyman, a Warmian canon was in practice a minor nobleman, holding a status similar to a knight in medieval England. Each one was required to keep weapons and to have at least two servants and three horses (the names of two of Copernicus's servants are known: Hieronim and Wojciech Szebulski). Most of the other fifteen canons with whom Copernicus resided were from the privileged classes of the Warmia region, and they had coveted the position they now held. They were all well educated and had attended a university; a university education was mandatory, in fact. Most were native German speakers, as was nearly everyone in this region of Poland, but in formal settings, they wrote and spoke in Latin.

Though the bishop of Warmia was the sovereign, and the king of Poland the ruler of the entire realm, the canons themselves actually controlled several sections of Warmia. So, in practice, the bishop and the canons were the joint rulers of the small kingdom. The canons governed the land around Frombork, and they also reigned over the region around the town of Olsztyn, about forty miles away.

Just like government officials today, most of their attention was devoted to the acquisition and distribution of money. Their main duty was to collect taxes, rents, and tithe from the peasantry who farmed and grazed the lands of the small kingdom and the artisans who worked and lived in the villages. After the bishop took his hefty share, which included a portion for Rome, the rest was divided among the canons. Many of the canons had other honorary positions, called sinecures or benefices, which allotted them even more financial resources. It was a very priviledged existence, and it is clear that Copernicus never wanted for income during his entire life.

In addition to collecting taxes, the chief obligations of the canons were to enact and administer laws, serve as judges, and appoint mayors, hunting inspectors, fishing inspectors, and other officials in the villages. The records from the era show the canons resolving disputes over fishing and hunting rights, awarding licenses for brewing beer, ruling on who could or could not carry arms, policing shipments of coins and jewels, and in one case trying to solve the kidnapping of a servant.

BECAUSE OF HIS MEDICAL TRAINING in Padua, Copernicus became the chapter's doctor when he moved to Frombork in 1510. The small town had its own hospital, the Hospital of the Holy Spirit, which was located down the hill but still quite close to the cathedral. It was here that Copernicus tended to the townspeople of Frombork, as well as his fellow canons. He probably spent some time at the hospital each day.

Because he would practice medicine for the rest of his life, he was usually referred to as Doctor Nicolaus. As a doctor, Copernicus was typical of the general practitioners of the day. He owned several medical books, most of which had been published in the late fifteenth century and which he probably had bought in Padua. He was not a surgeon, and there is no evidence that he was an active blood letter, either. Instead, he seemed to offer remedies of traditional concoctions. One recipe that survives in Copernicus's hand is probably reflective of how he and other nonsurgeons practiced their craft during the early sixteenth century. It was for something called the Imperial pill, which was evidently a catchall. It consisted of about twenty ingredients, among them gillyflower, agaric, seena, aloe syrup, anise, amomum, and the essences of violets and roses. The pill was a treatment for gray hair, poor vision, poor digestion, coughs, stomach gas, nerves, tooth decay, gout, insomnia, and colic.

Doctor Nicolaus also joined the other canons in taking turns holding the posts necessary to run their chapter and territories. Over the years, he served in each of the various positions, such as chancellor, treasurer, cantor, archdeacon, and custodian. In 1516, Copernicus assumed the important job of supervising the lands around the canon-governed town of Olsztyn. He left Frombork and moved into the castle there. Like the castle at Lidzbark, where he had lived with his uncle, the castle at Olsztyn was built by the Teutonic Knights and was massive. The doctor was now deeply immersed in his astronomical studies, and he was regularly making observations. Copernicus left behind evidence of his intense focus on astronomy while ruling Olsztyn and its environs—his etchings remain on the plaster walls of one of his studies.

Unfortunately, he did not have a great deal of free time while in Olsztyn. As the supervisor, he was constantly on call to deal with the problems of managing the land. There are dozens of records of Copernicus visiting the small villages and crossroads in this region and presiding over hearings in which land was taken from one resident and given to another. These official minutes are fascinating, but also depressing, as one reads about Copernicus giving parcels of land to new tenants when the previous tenant had become too old and senile, or had been crippled in a farm accident. There is never a mention about what happened to the unfortunates removed from the land.

Much more crucial than his role as land manager, Copernicus at one point had to become a military leader. After the Peace of Toruń in 1466, the Teutonic Knights held onto only one territory, but it essentially surrounded Warmia. When Copernicus joined his uncle's court in 1503, he was reminded of the Knights' continued presence. He and Bishop Lucas were constantly filing complaints with the king of Poland about marauders from the Knights' territory illegally crossing the border into Warmia, stealing horses and crops, and otherwise terrorizing the small farmers of their domain. By the time he took over as administrator of the lands around Olsztyn, Grand Master Albert and the Knights were itching to throw off the shackles of the fifty-year-old Peace of Toruń.

The Knights crossed the border into Warmia in December 1519 with 5,000 soldiers and cavalry. They seized, looted, and then burned many villages. In January of 1520, they attacked Frombork itself. They torched the entire town, destroying nearly every home, including the *curias* of the canons, Copernicus's among them. Only the cathedral was spared. The canons scattered to Gdansk and Elblag before the Knights could capture them.

After the torching of Frombork, the Knights occupied many parts of Warmia, but the aggression seems to have ended for most of that year. Then, toward the end of 1520, the Grand Master's troops were again on the move, threatening Olsztyn, where Copernicus was in charge. On January 16, 1521, the Knights demanded the surrender of Olsztyn, but Copernicus refused to yield. A small contingent of Polish calvary had arrived by then, and the Polish soldiers confronted the Knights in a skirmish, which surprised the enemy and delayed the attack for more than a week. But then the Knights assaulted Olsztyn on January 26 and broke through the first gate before being repulsed and forced to retreat. In February the Knights lifted the siege.

The hostilities ended several months later, when the Knights abruptly halted their campaign. The king of Poland threatened to send a large army to confront them, and Albert was running out of funds. Copernicus was then part of the small delegation that represented Warmia at the peace negotiations.

This was the last time that the Teutonic Knights would invade a territory, so for the citizens of Warmia the threat on their immediate border was finally over. The rest of the 1520s passed without any major disruptions in the bishopric, though the rebuilding of the towns, villages, and farms would take years to complete. Unfortunately, a new peril was already materializing, this one in Wittenberg, a couple of hundred miles away. Copernicus finished his term at Olsztyn and moved back to what remained of Frombork. He continued his ambitious astronomical research, making numerous observations, and his manuscript kept growing.

6

Before the Storm

On the day after Christmas in 1531, Bishop Maurice Ferber drafted a short letter to the canons at Frombork, urging them to send Nicolaus Copernicus to his palace in Lidzbark. Ferber was ill and wanted Copernicus, who was the doctor for the canons and the townspeople of Frombork, to attend to him. Knowing that he had no option but to comply, Copernicus instructed his servant to prepare the horses for the forty-mile trip. The father of astronomy may well have thought that the prospect of this tedious two-day journey in the winter cold and on the muddy, rutted roads was a fitting end to a year that he would rather forget. It had certainly been a rough one.

Ferber himself had been the cause of two deep low points. The worst was the letter that Copernicus received from the bishop in July reprimanding him for inappropriate behavior with his ex-housekeeper on her return from the Królewiec fair. The canon had been forced to respond to the mortifying letter, which must have pained him greatly (this is the letter cited at the beginning of the book). The unsavory exchange with the bishop revealed to Copernicus that one of his fellow canons was spying on him.

The other ignominious event had occurred earlier in the year. On February 4, apparently out of the blue, Ferber had threatened to cut off Copernicus's income if he did not complete the final steps necessary to become a priest. Since all the canons had already taken "first orders," it was not difficult to take "higher orders" and enter the priesthood. Copernicus had refused. We don't know why he declined; no explanatory document survives. There is no evidence that Ferber followed through on his threat.

Until this year, the fifty-eight-year-old Copernicus had been one of the most revered canons in the chapter, his record nothing short of exemplary. He had served his uncle, Bishop Lucas, as his assistant for seven busy years. He was the resident doctor. In 1516, he was assigned by his fellow canons to reside in Olsztyn castle and administer the chapter's lands surrounding it. He held that position for five years, the years that coincided with the last offensive of the Teutonic Knights. Copernicus—astronomer, mathematician, doctor—suddenly had to become a wartime leader and construct defenses against an army of calvary, infantry, siege-trains, and harquebuses. He had performed calmly and bravely while many of his fellow canons lived out these dangerous months in Gdansk. Once peace was restored, he finished his term in Olsztyn and returned to Frombork. There he helped dictate the final terms of the treaty with the Knights. During the 1520s, he spearheaded an attempted reform of the local currency intended to lessen the burden on the peasants, who had suffered horribly during that last martial gasp of the Knights. All in all, it was an admirable record of service.

Though no one would have known it by the nature of the personal attacks directed at him, Copernicus had known Bishop Ferber for decades. In fact, before Ferber became bishop and Copernicus's sovereign in 1524, the two men had been fellow canons at Frombork. They had lived together and probably interacted daily for years. What had happened to sour their relationship? It is possible that Copernicus just happened to be in

Ferber's line of fire because of the bishop's increasing anxiety over the movement now known as the Protestant Reformation, which Martin Luther had recently begun. He was in no mood to be dealing with rumors about his canons keeping mistresses and their refusal to join the priesthood. And, it was just as possible that Copernicus rebuffed the plea to become a priest for the reasons Ferber feared—that the astronomer knew he was attracted to women and had trouble keeping his vow of celibacy, and perhaps even that he thought some of the teachings of the Lutherans were worthy.

Ferber's anti-Protestant zealousness was stirred by a specific event in 1531, one that raised an alarm that the heretics were making inroads in the immediate region. The town of Elblag, the home of the ex-housekeeper, was again implicated, this time because Dutch followers of Luther and another reformer, Ulrich Zwingli, had recently settled there and were starting to proselytize. They were responsible for a play that was performed during the Fat Tuesday carnival in which the pope, cardinals, and bishops were defiantly ridiculed. Ferber was profoundly offended and worked tirelessly to have the players punished.

THOUGH THE DUTCH proselytizers had brought the Reformation directly to Warmia—Elblag was less than twenty miles from Frombork—the Reformation had been creating general tension for a long time.

The rupture that eventually became the Protestant Reformation officially started in the autumn of 1517, when Ferber was still a canon and Copernicus was dealing with the interloping Teutonic Knights. In October of that year, a monk named Martin Luther (1483–1546), a professor of theology at the University of Wittenberg, posted a list of abuses that he said were encouraged by the Church hierarchy in Rome. The list later came to be known as the Ninety-five Theses (each thesis was an abuse that Luther demanded must be discontinued) and it was hung on the main door of the massive Castle Church of Wittenberg. The specific event that drove Luther into action was the demand from Rome that all Christians contribute to the rebuilding of St. Peter's Basilica in Rome. To raise funds, the pope ordered the Dominican friars to fan out throughout Christendom and collect money. The specific tool that they used to separate the precious coins from the impoverished peasants and laborers was the selling of indulgences. An indulgence was a piece of parchment, the purchase of which subtracted a certain number of years from the amount of time a person would spend in Purgatory before entering the kingdom of heaven. One could also buy an indulgence to reduce the Purgatory sentence for a deceased relative or for one who was still alive. Most of Luther's theses concerned the indulgences.

Though the posting of the Ninety-five Theses was later recognized as the spark that lit the fire, this was hardly Luther's intention. He did not set out to found a new religion or to cause a profound rupture in the Church. But he did want to see major changes in the Church, and he was an active and impassioned champion for those changes. Also, as time went on and his support increased, he became emboldened—for instance, in 1519, he formally denied the supreme authority of the pope (the Leipzig Disputation).

Pope Leo X excommunicated Luther in June 1520. In order to defend himself before the sentence was executed, Luther was ordered to appear in April 1521 at the Diet of Worms to recant his heresies. To the shock of the officials gathered there, not only did Luther not recant, he defiantly burned the parchment that contained the papal bull of excommunication. The pope's condemnation immediately became official and Emperor Charles V outlawed Luther and banned his writings from the Holy Roman Empire. Luther fled from Worms and went into hiding under the protection of the Elector of Saxony at his castle in Wartburg. During the ensuing months, word came to the castle that there was a surprising amount of support for Luther's position and admiration for his brave actions. Heartened, he returned to Wittenberg in March 1522. The town leaders immediately embraced Luther and endorsed his reforms, thus becoming the first government to break from the Church. The Reformation was officially under way.

The initial act of the reformers was to publish the New Testament in vernacular German. The first installment appeared in 1523 (the first installment of the Old Testament was released later that year), as they stressed that all Christians, not just the clergy, must know the word of God.

The boldness of Luther and the citizens of Wittenberg opened the floodgates, and long pent-up complaints against the Roman Church were voiced all over central Europe. Other reformers and their followers materialized almost overnight, in particular Ulrich Zwingli in Switzerland in 1522. There were excesses, too. Inspired by Luther, the peasants of south Germany rose against the Catholic clergy and the landowners. The brutality of the peasants was matched only by that of the forces arrayed to beat them back into submission. After the Peasants' War (1524–1525), as the rebellion was later called, there was no going back. During the war, the city of Nuremberg, one of the largest cities in Europe, officially adopted Lutheranism (1524).

Numerous free cities and principalities in the German territories elected to tie their fortunes to the reformers. The first major country to adopt Lutheranism was Sweden in 1529. The king of Denmark also converted in 1529, forcing his country to become Lutheran, too. Many more states followed in quick succession.

For the bishop, clergy, and citizens of Warmia, the Reformation manifested itself early. After the hostilities between the Teutonic Knights and Warmia (allied with the Kingdom of Poland) ended, but while the peace was very fragile, Grand Master Albert traveled to Nuremberg in 1522. While there he met Andreas Osiander, a Lutheran reformer who will later play a very different role in this story. Osiander was a zealot for the Lutheran cause and he converted Albert to Lutheranism, though the conversion would not be official for a few years. Albert returned to his castle in Królewiec and reluctantly decided that he could not afford a resumption of hostilities against Warmia and Poland. Thus he sought a final peace. The Treaty of Kraków was signed in April 1525. The agreement forced Albert and the Knights to come more firmly under the control of Poland, which was welcome news in Lidzbark, Olsztyn, and Frombork. Just four months later, though, Albert announced that he was a Lutheran. In December he issued a Church Ordinance that converted his principality to Lutheranism. This was *not* welcome news in the Warmian towns. Teutonic Prussia now became Ducal Prussia, and Grand Master Albert was now the Duke of Prussia. This meant that Warmia's long-time nemesis—a territory that nearly surrounded the small bishopric—was the first government larger than a city to declare itself officially Lutheran. Fortunately for Warmia, Ducal Prussia was now so impoverished that it never seriously threatened the bishopric again. Still, the conversion of Albert was alarming for Bishop Ferber and the King of Poland.

That same year, 1525, riots among the citizens of Gdansk forced the city council to accept Lutheranism. The king of

Poland would not stand for this, so he formed an army of 8,000 soldiers and personally led an expedition in the summer of 1526 to remove the reformers from Gdansk, the most important city in the northern part of the kingdom. King Sigismund's expedition forced the reformers to flee, but there was religious tension in Gdansk for decades afterward.

The events in Ducal Prussia and Gdansk did not go unnoticed in Warmia. In September 1526, Ferber issued an order expelling all Lutherans from Warmia, and demanding that all pro-Lutheran publications be confiscated and destroyed. Anxiety was building again in 1531—not only were the heretics active in Elblag, but the new Protestant (the term came into use in 1529) lands formed an alliance that year against the Holy Roman Empire, which had remained staunchly Catholic. The formation of the Schmalkaldic League, which the alliance was called, made a religious war a definite possibility. All of Europe was apprehensive.

ON THE ROAD TO LIDZBARK, Copernicus and his servant rode their horses side by side as they made slow progress across the snowy countryside. The horses needed little directing since they traveled the route all the time, at least several times a year and sometimes more. The servant may have remarked to himself how distinguished-looking his master was. Though now an older man, Copernicus was probably still considered handsome, or if not handsome, certainly striking. He had a long, thin face, with high cheekbones, a strong jaw, and a pronounced chin. The facial feature that dominated was his thin, but large, nose. Picture the English actor Basil Rathbone, or perhaps James Cromwell (from *L.A. Confidential* and *The Queen*) and you have an image of what Copernicus looked like. (This description is based on

Copernicus's self-portrait and a forensic reconstruction of his skull, which was rediscovered in 2005. If that skull is really Copernicus's, and if the forensic reconstruction is accurate, then his nose had been broken at least once, and he had a scar above his right eye, which drooped somewhat.) It is believed that Copernicus still had his hair as an adult and that he wore it somewhat long and shaggy. As a young man he was clean-shaven and quite handsome, as reflected in his self-portrait, which he painted in the early 1500s. Apparently, good looks ran in his family, as his maternal grandmother, Katherine Watzenrode was considered "the pearl of all Toruń beauties."

Though the servant lived with Copernicus in his *curia*, he probably did not know his master intimately. Not because the doctor treated his help indifferently, but because he was known to be a quiet, introspective man. No one really knew Copernicus that well. The most reputable reference to Copernicus's personality comes from a letter written a dozen years later, when Copernicus was on his deathbed, from his best friend, Tiedemann Giese, who from a distance was trying to ensure that someone would stand by Copernicus's side and comfort him: "Just as he loved privacy while his constitution was sound, so, I think now that he is sick, there are few friends who are affected by his condition." Giese was deeply worried that Copernicus would die alone.

It cannot be known for certain, but several personal losses in his life may have contributed to Copernicus's distant personality. His father passed away when he was only about ten years old. His protector and benefactor, Uncle Lucas, died two years after Nicolaus resigned as his personal secretary. Lucas Watzenrode was a very powerful man with many enemies, and he died so suddenly in 1512 that some suspected foul play.

His worst loss, though, was the death of his brother Andreas in 1518. As discussed earlier, Andreas was probably about two years older than Nicolaus, and the Copernicus brothers were quite close. They essentially had led parallel lives once they left Toruń for the University of Kraków in 1491. Though there may have been some tension between them because Uncle Lucas consistently took care of Nicolaus first—after both young men finished at the University of Kraków, Nicolaus went to the University of Bologna two years ahead of Andreas, and Nicolaus also became a canon at Frombork two years earlier, too—there is no evidence of any animosity. The two times that the brothers appear together in the documentary record, they seem close. In the fall of 1499, the brothers needed to borrow money together—"Your Reverence's nephews, who live in Bologna, were short of money these past days, in the manner of students. They betook themselves to George [Uncle Lucas's secretary, who was visiting Bologna] as truly destitute to someone destitute, asking him what advice he had. Andreas proposed to seek employment in Rome to alleviate his poverty. They finally received 100 ducats from a bank, charging interest." Then, "In the year 1501, before the chapter there appeared Canons Nicolaus and Andreas Copernicus, who are brothers . . . After thorough consideration, the chapter acquiesced in the wishes of both brothers." Their wishes were to return to Italy and continue their studies.

When the young men departed Italy for Warmia in 1503, Andreas left with the worst possible souvenir—a debilitating, ultimately fatal infectious disease. The chapter records called it *lepra*. It could have been leprosy, but it also could have been syphilis (leprosy was becoming rarer at this time, while syphilis was increasingly common, thanks to Columbus and his sailors returning from the New World). Both diseases are painful and disfiguring, and take years to run their course. The chapter records show that Andreas was considered "*contagioso*," so his affliction may have been leprosy. The symptoms are the eruption of ugly, painful lesions that appear on the skin and that often lead to deformed ears, fingers, and toes. But worse is what leprosy does to the victim's face: The nose sinks into the nasal cavity because bacteria eats away the nasal bone. Andreas was well enough for a couple of years, but by 1508 he felt the need to leave Frombork and return to Italy for treatment. He came back to Frombork in 1512, but the disease had advanced and his presence was no longer wanted. At a meeting in September 1512, Andreas was asked to leave the chapter, after first returning the large sum of 1,200 gold Hungarian florins (this debt and the fact that he had not paid his tuition in full back in 1491 perhaps reflected his undisciplined ways). Nicolaus was present when the vote was taken to expel Andreas from Frombork. Andreas eventually did leave Frombork for Rome, where he probably died in 1518.

The deaths, most especially his brother's, seemed to have darkened Nicolaus's demeanor.

THOUGH THE SITUATION with the former housekeeper indicates that Copernicus may have acted younger than his nearly sixty years in matters of the heart, it appears that in other ways he was winding down his career. In fact, the troubles with Ferber coincided with the canon's deliberate decision to distance himself from his previous duties. It may have been that at age fifty-eight he felt that it was time for the next generation to take the stage. Or maybe there were simply too many embarassing events occurring and he decided to hide in his shell. Whatever the precise reasons, by the end of 1531, the old spark was flickering away, and Copernicus went into what today we would call semiretirement. After years of being in the thick of religious, economic, and political events, and having been deeply involved in wars and church intrigue, he now let others take the active roles.

Copernicus would never again be as active a canon as he had been prior to this time. At one point he was forced to give up two of the other benefices that he had held for decades. Most distressing of all, on February 19, 1535, Copernicus's sixty-second birthday, Ferber wrote to the canon to remind him to choose a successor for his canonship. There is no denying that your days are numbered when you are asked to plan for your succession.

IRONICALLY, THE 1530S brought a certain amount of recognition to Copernicus's astronomical work, as events outside of Frombork and Warmia were adding to his acclaim. So, while Canon Nicolaus was made to feel uncomfortable and marginalized at home, astronomer Copernicus, unbeknownst to him, was achieving recognition.

In 1535, Bernard Wapowski, Copernicus's University of Kraków friend, wrote a letter to a gentleman in Vienna urging him to publish an enclosed almanac, which he claimed was written by Copernicus. This is the first and only mention of a Copernicus almanac in the historical records, but this official missive is difficult to dismiss. More likely, the almanac was Copernicus's tables of planetary positions, which would have been useful for calendars and astrology. The Wapowski letter mentions Copernicus's theory about the motions of the earth, so the old acquaintance definitely knew about Copernicus's heliocentric theory. Nothing came of the directive because Wapowski died just a couple of weeks later.

More significant is what was going on in Rome. A secretary of the pope, a gentleman named Johann Widmanstetter, explained the Copernican system to Pope Clement VII and two cardinals in the summer of 1533. The pope was so pleased that he gave Widmanstetter a valuable gift. After Pope Clement's death, Widmanstetter took a position with Cardinal Nicholas Schönberg, and through Widmanstetter the cardinal, too, learned of Copernicus's theory. Perhaps with the achievements of his distant predecessor Bessarion in mind, Schönberg sent the canon a most remarkable letter, dated November 1, 1536:

> Some years ago, word reached me concerning your proficiency, of which everybody constantly spoke. At that time, I began to have a very high regard for you, and also to congratulate our contemporaries among whom you enjoyed such great prestige. For I had learned that you had not merely mastered the discoveries of the ancient astronomers uncommonly well but had also formulated a new cosmology. In it you maintain that the earth moves, that the sun occupies the lowest, and thus the central place, in the universe . . . I have also learned that you have written an exposition of this whole system of astronomy, and have computed the planetary motions and set them down in tables, to the greatest admiration of all. Therefore with the utmost earnestness I entreat you, most learned sir, unless I inconvenience you, to communicate this discovery of yours to scholars . . . I have instructed Theodoric of Reden to have everything copied in your quarters at my expense and dispatched to me. If you gratify my desire in this matter, you will see that you

> are dealing with a man who is zealous for your reputation and eager to do justice to so fine a talent.

This letter shows that Copernicus's ideas were being discussed in the early 1530s as far away as Rome, and that the long manuscript was largely finished by 1536. But Copernicus did not make his manuscript available to Schönberg. He chose to ignore the cardinal's earnest plea.

Why did he disregard this generous offer? By this time Copernicus was troubled about the fact that he had not been able to prove parts of his heliocentric theory to his complete satisfaction. Perhaps for this reason, perhaps for others that we don't know, Copernicus decided to share his manuscript with only a few intimates and not to circulate it more widely.

7

The Death of the Bishop

The canons at the chapter in Frombork issued the following official announcement on July 1, 1537:

> The Chapter of Warmia notifies through a mandate to the Bishop's officeholders that, because of the death of Maurice Ferber, Bishop of Warmia, on July 1, 1537, it has taken over complete authority for the Warmian lands until the election of his successor. It has delegated to Lidzbark as its plenipotentiary administrators and legal advisors Nicolaus Copernicus and Felix Reich, Canons of Warmia, to prepare there a list of the late Bishop's things, the things belonging to the Bishopric, and for them to take care of the affairs of the castles, towns, office-holders and subjects of the Bishop. The Chapter calls for the manifestation of obedience to its delegates.

Maurice Ferber had finally passed away after several years of failing health that had caused him to miss many official functions. Doctor Nicolaus had visited him numerous times to help alleviate his pain and discomfort. Now he was once again en route to Lidzbark, this time as an official representative of the chapter itself. With the passing of the bishop, the chapter of canons was now the collective leader of Warmia until the new bishop was elected. Surprisingly, given Ferber's serious final illness, critical documents, artifacts, and especially coffers filled with coins and valuables had been left practically unattended in Lidzbark Palace. That was why Copernicus and fellow canon Felix Reich were sent to the palace in such haste. They were the two oldest canons and had probably been chosen because they were the most senior, and therefore most distinguished, representatives of the chapter. Plus, Copernicus understood better than anyone the intrigues that might exist at the palace, given the fact that he had lived there with his uncle Lucas three decades earlier.

After a night at an inn, Copernicus's and Reich's coaches would have pulled into the stone courtyard of Lidzbark Palace sometime on July 2. Fortunately, all was in good order and the two canons were able to secure the treasury, all important documents, and the body of Ferber expeditiously. On July 4 or 5, they loaded their two coaches with the valuables and the deceased, and along with their servants began the return trip to Frombork.

Several days later, Bishop Ferber was given an appropriate funeral in the beautiful cathedral of Frombork. His body was interred in the cathedral itself, joining most of the the past bishops of Warmia, including Copernicus's uncle Lucas.

THE SAME DAY that Copernicus began his trip to Lidzbark, two letters bearing news of Ferber's death were sent by two different canons at Frombork to Bishop Johannes Dantiscus, the ruler of the neighboring bishopric of Chelmno. (This is the same Chelmno that was the site of one Hitler's extermination camps; Chelmno was the name for both a bishopric and a town.) The two canons were eager to inform Dantiscus about the passing of Ferber because it was essentially a fait accompli that Dantiscus would be his successor. Chelmno was much smaller than Warmia, and it was known that the bishop of Chelmno was going to be rewarded for outstanding service to the king of Poland and the pope by ascending to the post of bishop of Warmia when it became vacant. In fact, Dantiscus had bragged in a letter several months before Ferber's passing that he was slated to take over Warmia, which was four times wealthier than the bishopric of Chelmno.

There was no reason to suspect that this change at the top would be problematic for Copernicus. Dantiscus had been the bishop of Chelmno since late 1532, and had been made a canon at Frombork three years earlier, in 1529 (though it is not certain how much time, if any, Dantiscus spent residing in Frombork). Dantiscus was twelve years younger than the astronomer (he was born in 1485), but had known him for a long time. They first interacted in the early years of the century, specifically 1507–1510, when Copernicus served his uncle Lucas and Dantiscus was the ambassador of the king of Poland, Sigismund I, to the Prussian Estates. An extraordinarily well educated and informed man, Dantiscus was probably one of the few individuals in the region who knew about Copernicus's talents as a mathematician, and he likely even had some understanding about the revolutionary nature of his astronomical theories. Because he was such an insider, he was certainly aware of Cardinal Schönberg's invitation to publish Copernicus's book. In addition, Dantiscus had recently become very close to Tiedemann Giese, another canon at Frombork who was Copernicus's best friend. The Dantiscus-Giese relationship blossomed because Giese had been chosen to be Dantiscus's successor as bishop of Chelmno. As Ferber's health declined in the early months of 1537, there was a great deal of communication between Dantiscus and Giese concerning the transfer of positions that would occur shortly.

Dantiscus was born Johannes Flachsbinder, the son of a German brewer and his wife, in Gdansk in 1485. The senior Flachsbinder must have been a very successful brewer, because he was able to send his son to the distinguished University of Kraków, starting in 1500, when the boy was fifteen. Of course, Copernicus also attended the University of Kraków, but a decade earlier. As many did, Flachsbinder latinized his name to Dantiscus (after his hometown of Gdansk) when he entered the university. While still a student, he joined the army of the king of Poland and fought in the campaign against the Tatars and Wallachians in 1502. Resuming his studies, Dantiscus graduated in 1503, and then joined the court of the Polish king Aleksander Jagiellon as a scribe in the royal chancellery. A restless youth, he embarked on a pilgrimage to the Holy Land at the age of twenty, where he stayed for two years. The same spiritual urge that motivated him to make the pilgrimage would later manifest itself as piety and strict adherence to the rules of the church . . . but not as a young man. On his return from Jerusalem, the brewer's son once again found employ with the king of Poland, this time with the young Sigismund I, who had succeeded his father on the elder's death in 1506. Dantiscus would maintain a strong connection to the king, and to his two successive wives, for the rest of his life. The young man from Gdansk became the secretary to the king, and shortly thereafter his special envoy to the Prussian Estates. In addition to his official duties, Dantiscus was also a noted and prolific poet who would publish many volumes of verse in the years ahead.

Dantiscus had achieved all this by the age of twenty-three. By contrast, at the same age, Copernicus was still a student in Italy. After their joint activity in the Prussian Estates, the two

University of Kraków alumni went in very different directions. Copernicus left his uncle in Lidzbark and settled in Frombork as a canon in 1510. Dantiscus became the official ambassador of the king of Poland to the powerful courts of the Holy Roman emperors Maximilian I and Charles V, and to the court of the king of Spain, Charles I. He was so popular with these monarchs that Maximilian I made him poet laureate, in recognition of his gift with verse, and Charles I gave him a landed title in Spain. He also served as a special envoy for both emperors during various delicate negotiations elsewhere in Europe. During these years of high activity, he formed personal relationships with King Henry VIII of England, Cardinal Wolsey, the king of Denmark, and the regent of the Netherlands.

As a scholar once observed, Dantiscus's main interests during his ambassador years "had been poetry, women, and the company of learned men, apparently in that order." He had one of the more extensive correspondences of the period. One of his correspondents was a noted mathematician named Gemma Frisius, and there is an exchange of letters in which they discuss the motion of the earth and the heavens.

Although in his portrait he appeared to be rotund, Dantiscus was known as a notorious womanizer. Two of his mistresses are known by name: Grinea, who lived in Innsbruck, and Isabel Delgada of Toledo, with whom he had a daughter, Dantisca.

In the late 1520s, when he was in his early forties, Dantiscus decided to leave his ambassador life behind and join the clergy in the region of his birth. Although his decision to step off the world stage surprised his many acquaintances, it may have been planned for years. In 1514, Dantiscus had attempted to become the successor to Copernicus's brother, Andreas, who was already ill. To become bishop of Warmia, the only requirement was that one had to already be a canon at

Frombork. In 1529, Dantiscus finally did secure a canonry, and only a year later, May 4, 1530, he was named the bishop of Chelmno (he officially assumed the position in September 1532), through the influence of King Sigismund I. As bishop of Chelmno, he was virtually guaranteed to succeed Ferber as the bishop of the much bigger and richer bishopric of Warmia.

SHORTLY AFTER BEING ELECTED bishop of Chelmno, Dantiscus sent Copernicus a very complimentary letter, inviting the canon to visit him at his castle in the town of Lubawa, possibly for the celebration of his first official mass as the new bishop. Dantiscus also invited Felix Reich, so he may have just been recognizing the two senior canons whose support he would someday need when ascending to the bishopric of Warmia. He may also have wanted to know more about the distinguished intellect in his midst. However, Copernicus declined the invitation, writing:

> I have received Your Most Reverend Lordship's letter and I understand well enough Your Lordship's grace and good will towards me; which he has condescended to extend not only to me, but to other men of great excellence. It is, I believe, certainly to be attributed not to my merits, but to the well-known goodness of Your Rev. Lordship. Would that some time I should be able to deserve these things. I certainly rejoice, more than can be said, to have found such a Lord and Patron. However, regarding Your Rev. Lordship's invitation to join him on the 20th of this month (and that I should most willing do, having no little cause to visit so great a friend and

> patron), misfortune prevents me from doing so, as at that time certain business matters and necessary occasions compel both Master Felix and me to remain at this place . . .

There is no document recording further interactions until three years later, when in June 1536, Copernicus wrote to Dantiscus to acknowledge, but again decline, an invitation to Lubawa, this time to attend the wedding of one of Dantiscus's relatives: "Truly, Your Rev. Lordship, I ought to obey Your Lordship and present myself from time to time to so great a Lord and Patron . . ." This reply suggests that there were other entreaties from Dantiscus urging Copernicus to join him at his castle in Lubawa.

There was one additional letter from Copernicus to Dantiscus before he became the Bishop of Warmia. This time it appears that Copernicus initiated the contact, and it was a bland letter containing news from abroad. Copernicus may simply have been trying to ingratiate himself with his future superior.

THERE WERE SEVERAL PROTOCOLS to conduct before Dantiscus could formally ascend to the position of bishop of Warmia. The most important was the official election by the sixteen canons of the chapter at Frombork. First, the canons produced a slate of four candidates that had to be approved by King Sigismund. Of course, Dantiscus was one nominee; Copernicus was one of the other three, out of courtesy.

Sigismund sent a letter to the canons at Frombork on September 4, 1537, approving the four nominees and also expressing confidence that the first of the candidates, Dantiscus, would win the election. The election itself was made official on September 20, when the chapter sent word to King Sigismund that Paul Plotkowski, provost; Leonard Niederhoff, cantor; Tiedemann Giese, custodian and plenipotentiary of Johannes Dantiscus; Johannes Zimmerman, cantor; Alexander Scultetus; Felix Reich; Paul Snopek; Nicolaus Copernicus; and Achacy von Trenck had unanimously elected Johannes Dantiscus as the new bishop of Warmia.

Now that Dantiscus was bishop, there was bound to be some anxiety on the part of the canons. After all, Ferber had been their sovereign for fourteen years and they were accustomed to his ways. But Copernicus had little to fret about, though he would miss his good friend Tiedemann Giese, who was now succeeding Dantiscus as bishop of Chelmno. In fact, during April 1538, Dantiscus felt that he needed medical attention and he requested that Copernicus stay at the palace in Lidzbark until he felt better. This time Copernicus complied and stayed in Lidzbark for several weeks. Dantiscus later wrote to Giese to say "I feel better, as the doctor, our respected and common friend, will tell you in greater detail. His gentle manner and conversation, and his advice, as soon as I took it, were a cure for me." Dantiscus even asked Copernicus to effectively handle the election of the new canon to fill the vacancy created by Giese's elevation to bishop.

By the beginning of the summer of 1538, the transition to the new bishop was nearly complete. Copernicus was getting to know Dantiscus, and, thanks to the canon's skills as a doctor, they were already forming a warm relationship. There was only one more element of protocol that remained: the official tour of the kingdom by the new bishop.

8

The Mistress and the Frombork Wenches

ALMOST PRECISELY one year after he had boarded his coach to make the solemn journey to Lidzbark to collect the body of the deceased Maurice Ferber, Copernicus, accompanied by Felix Reich, was returning to Lidzbark, this time to celebrate.

Johannes Dantiscus was ready to perform the final ritual required to seal his ascension to the bishopric of Warmia—an official tour of a select number of towns in the realm. At each stop the merchants and artisans, the burgers and gentry, the various town officials, and the peasants from the surrounding countryside, were to display their respect for the new sovereign by solemnly reciting an oath of allegiance. Two canons from the chapter that had formally elected him the previous September were to accompany him. Copernicus and Reich had been chosen, probably for the same reason that they were dispatched to secure the body of Ferber the year before—they were the two most senior canons in the chapter and thus its most visible representatives. This would be their last joint task.

The two aging canons joined Dantiscus at the palace in Lidzbark toward the end of July 1538. The oath-of-allegiance tour started there, and then continued through Reszel, Jeriorany, Barczewo, Dobre Miasto, and Orneta, concluding in Braniewo, and then circled back to Lidzbark. The selected towns had once been strongholds of the Teutonic Knights, so each had a cathedral for the ceremonies and a castle for the entourage to reside in during the evening.

The three clerics spent many hours in one another's company as they traveled from town to town. During the daylight hours they were either on horseback together as they progressed slowly on the dusty roads in the muggy summer heat, or else they were in one of the Warmian towns. When the sun started to set, which was late in this northern region in midsummer, they retired to the local castle, which had been prepared for their arrival. One can imagine the conversations being somewhat more animated and less guarded in the evening, as the beer and wine flowed freely. (Though there are no references in the documentary record of Copernicus being a drinker, we know for certain that Canon Felix enjoyed both wine and beer, and that Bishop Dantiscus enjoyed wine. Copernicus must have drank beer and wine, too, as did everyone in the sixteenth century, because only fermented beverages could be imbibed safely.) Also, with Copernicus and Reich both getting far along in years and gradually retiring from their day-to-day duties as canons, they were probably less circumspect than they would have been earlier in their careers. Many topics must have been discussed, both official and personal.

These three older men, who had seen an amazing amount in their full lives, had plenty to talk about. Dantiscus, in particular, had much on his mind. At the very time when he should have been elated for having attained a revered title—impressive for the son of a brewer—he was deeply troubled by nearly everything transpiring around him. He even mentioned to a correspondent some months earlier that he was starting to believe that world events were portending the fulfillment of the biblical prophesies about the end of the world.

World events could not have been more disturbing and disheartening. In Poland, his friend King Sigismund I was fighting an ill-advised war against a Moldavian warlord (later called the Hen War), which was not going well. Worse, the power-hungry king of France, Francis I, a Catholic, had recently made an alliance with Suleiman I, the leader of the murderous, infidel Ottoman Turks, giving the Ottomans the potential to someday take over all of Europe. And, worst of all, the Lutheran Reformation was turning the world upside down. Kingdom after kingdom, province after province, city-state after city-state, free city after free city—it seemed that all of northern Europe was embracing the heretic Lutherans. Even Henry VIII of England, whom Dantiscus knew personally, officially severed ties with the Church in 1534, and two years later took the unthinkable step of executing his wife, Anne Boleyn. Soon after that appalling act, Henry abolished the monasteries in England. The king of Denmark had converted to Lutheranism and exiled the Catholic clergy. In fact, there were three Danish bishops who had escaped to Gdansk whom Dantiscus was personally looking after. Most ominously, though, were the inroads that Lutheranism was making within and around little Warmia. The Prussian noblemen had started to realize that removing distant Rome from their concerns would save them considerable amounts of wealth and alleviate many meddlesome intrusions. If all of the land near Warmia came under the spell of Luther, how stable would the bishopric remain, isolated and surrounded by its enemies in northern Poland?

Personal issues were also weighing heavily on Dantiscus. He was already embroiled in conflicts with certain canons at Frombork. Furthermore, he was upset with the amount of money he owed Rome for having attained the bishopric. In fact, Dantiscus devoted a great deal of ink in his correspondence to his personal finances; he even fussed over issues as small as postage costs. Perhaps because he came from modest roots, he seems to have had a self-made man's obsession with making sure his income was accounted for. While he was alive, Bishop Ferber had at least once felt compelled to chide Dantiscus in writing for complaining too much about his finances.

Most disturbing of all was the situation surrounding his daughter. Dantiscus's last posting in the service of the king of Poland had been in Spain, where he spent nine years at the court of Charles I. While there he met Isabel Delgada, a Spanish beauty with whom he had two children. One died in infancy, but the other, Juana Dantisca, born in 1527, was dear to him. Dantiscus never married Delgada, so his daughter was technically illegitimate, but her name made it clear that he was not trying to hide from his paternal responsibility. He regularly sent money to both of them. What was keeping him up at night in the summer of 1538, though, was his recent discovery that Isabel had allowed the betrothal of Juana to Diego Gracian de Alderete, a Spanish imperial secretary whom Dantiscus had known well while in Spain. He knew him too well, in fact. The groom-to-be was forty-three years old—thirty-two years older than the bishop's prepubescent eleven-year-old daughter. (However, it appears that if Dantiscus had been willing to part with 200 ducats, Isabel would have allowed Juana to join him in Warmia, presumably unmarried. Dantiscus would not send the ducats.) His letters to friends and to both Isabel and Alderete show that he was beside himself over the impending marriage.

One can picture Dantiscus talking to Copernius and Reich about his daughter one night and then Copernicus letting his guard down and mentioning his secret. Or perhaps it was Reich who inadvertently revealed what was common knowledge in Frombork—Doctor Nicolaus had a mistress who regularly visited his house. Copernicus would soon regret that this exchange ever happened.

Sometime between July 1531 and the summer of 1538, Copernicus started receiving a woman at his *curia* in Frombork. In the sixteenth century, the rules of celibacy among the clergy were often ignored and many priests and clerics lived with women, some even had families with them. Though it certainly occurred, the practice was by no means universal, and in Copernicus's circle of canons at Frombork, only three of the sixteen had mistresses—Copernicus, Leonard Niederhoff, and Alexander Scultetus. Scultetus was the only one who had children with his partner.

Though there is no way of knowing precisely what transpired, it is certain that Dantiscus learned about Copernicus's mistress during the tour. It is also certain that he was not at all sympathetic about the situation. Dantiscus had turned over a new leaf since becoming a cleric and took his vow of celibacy seriously. It is possible that the birth of his daughter had caused him to change his ways, for he left Spain and became a canon shortly after Juana's birth. In a letter drafted sixteen months before he went on the tour with Copernicus and Reich, Dantiscus cautioned one of his good friends who was interested in a career in the Church to change his habits and practice sexual restraint. He quoted Saint Paul and warned that "a man sullied by debauchery receives the sacrament of the Eucharist in an unworthy way." He expected the same adherence to Church dictates from his fellow clerics. Also, with the Lutherans vigorously accusing the Catholic clergy of corruption, flagrant flaunting of the rules could not be tolerated. Dantiscus told Copernicus that he must immediately end the relationship.

The woman was named Anna Schilling, and it is believed that she was the daughter of a Dutchman, Arend van der Shelling, who had settled in Gdansk. Van der Shelling married a distant relative of Copernicus's, and at that time he changed his name to Schilling. Copernicus and Anna's father were involved together for seven years, from 1529 to 1536, as the legal guardians of a group of orphaned children. It was most likely this connection that led Nicolaus and Anna to meet. She had her own house in Frombork. Most problematic for Copernicus was the fact that she was still technically married (we do not know anything about her husband), though presumably separated. Anna was reported to be pretty, well educated, and deeply interested in astronomy. She must have been more than just pretty, though, and blessed with a magnetic personality. This assumption is based on an exchange of letters after Copernicus's death in 1543. The canons of the Frombork chapter wrote to Dantiscus to ask if Anna could be allowed to move back to Frombork, from which she had been exiled, now that Doctor Nicolaus had passed away. Dantiscus wrote back—immediately—to say no, that Anna was a temptress and seducer and that she could ensnare other canons if allowed back to Frombork. His precise words were, "For it must be feared that by the methods by which she deranged him [Copernicus] . . . she may take hold of another of you, my brothers . . . I would consider it better to keep at a rather great distance than to let in the contagion of such a disease. How much she has harmed our church is not unknown to you, my brothers, for whom I hope happiness and health."

DESPITE DANTISCUS'S ORDER to clean his household, Copernicus did not comply when he returned to Frombork in August. So in November of 1538, Dantiscus felt compelled to remind Copernicus in writing that he had promised to remove

Anna from his premises. The reminder spurred this response from Copernicus to Dantiscus:

> My lord, Most Reverend Father in Christ, most gracious lord, to be heeded by me in everything:
>
> I acknowledge your Most Reverend Lordship's quite fatherly, and more than fatherly admonition, which I have felt even in my innermost being. I have not in the least forgotten the earlier one, which your Most Reverend Lordship delivered in person and in general. Although I wanted to do what you advised, nonetheless it was not easy to find a proper female relative forthwith, and therefore I intended to terminate this matter by the time of the Easter holidays. Now, however, lest your Most Reverend Lordship suppose that I am looking for an excuse to procrastinate, I have shortened the period to a month, that is, to the Christmas holidays, since it could not be shorter, as your Most Reverend Lordship may realize. For as far as I can, I want to avoid offending all good people, and still less your Most Reverend Lordship. To you, who have deserved my reverence, respect, and affection in the highest degree, I devote myself with all my faculties.
>
> Frombork, 2 December 1538
>
> Your Most Reverend Lordship's
> Most obedient
> Nicolaus Copernicus

Most of the canons had housekeepers, who were often relatives. Though everyone knew the actual situation between Copernicus and Anna, it appears that the astronomer told people that she was simply his housekeeper. So, it was difficult to throw her out of his house right away, because "it was not easy to find a proper female relative forthwith."

Six weeks later the situation seemed to have been addressed when Copernicus wrote to Dantiscus to tell him that "I have done what I neither would nor could have left undone, whereby I hope to have given satisfaction to Your Rev. Lordship's warning."

However, Copernicus had lied. He had not ended the relationship, and word of the continued presence of Anna Schilling in Copernicus's house got back to Dantiscus later that winter. Even before Dantiscus learned of Copernicus's deception, he had decided that he needed to apply more pressure, and that pressure had to come from within the Cathedral Hill walls. Clearly, the reprimands from Dantiscus's pen emanating from the distant palace of Lidzbark were not working. He chose as his Frombork ally the canon who had probably witnessed the first discussion between the bishop and Copernicus about Anna, Felix Reich. Canon Felix could not have been pleased to be placed in such an uncomfortable spot.

Through the winter of 1538–39 there was a comical exchange of letters between Dantiscus and Reich. The bishop wanted Reich to confront the doctor about Anna Schilling in front of the entire chapter, thus making it more personal and more difficult for Copernicus to ignore. Specifically, Dantiscus wanted Reich to read an official writ of misconduct.

But Reich not only accompanied Copernicus on numerous joint projects over their many years together, he was clearly fond of his colleague. Felix remembered Nicolaus in his will (he was one of the few cited); he readily acknowledged that Copernicus had successfully treated him for several illnesses, and he referred to him in one letter as "the venerable Nicolaus Copernicus." Reich wanted nothing to do with the unsavory task of reading the writ. So he conducted a delaying tactic. The first draft of the writ was sent back to Dantiscus because of a technicality:

> I hope that he [Copernicus] will take it [Dantiscus's warning] to heart, so as not to need my admonition. He will be overcome with shame, I am afraid, if he learns that I am privy to this matter. Had I not been prevented by the insertion of certain little words, I would perhaps have read to him your Lordship's letter insofar as it touches on that business.

Six weeks later, Reich again wrote to Dantiscus about Copernicus and the other two cohabiting canons, Scultetus and Niederhoff, this time appearing to be ready to help. He offered Dantiscus advice on how to properly word and deliver the official writs:

> God Almighty will strengthen your arm so that you may conduct to a happy ending what you initiated out of zeal. As much as we can, all of us will help make a success of this affair. However [there was always a "however" with Reich], your Most Reverend Lordship must take care nevertheless in commencing the proceeding with the force of law not to introduce in your future letters anything contrary to formal and customary legal style, as it is called. For it often happens that even the tiniest

> clause may spoil an entire case, so that it is declared null and void if it comes before a higher judge.

Dantiscus and his staff went right to work and quickly prepared the necessary documents. They sent them to Reich in Frombork immediately. But there was something wrong again: "I am sending back all of the letters because in one a serious scribal error must be corrected, and that cannot be done here. For the scribe wrote 'Henry' instead of 'Alexander' [Scultetus]"

In the final move of this dance, Felix Reich sent the revised letters back to Dantiscus on January 27, 1539, on the pretext that there were currently not enough canons in residence to conduct a formal meeting. Of course Reich could have held on to the letters until there were enough clerics in Frombork, but he did not even want them on his person at the cathedral.

Then, one month later, on March 1, 1539, Canon Felix Reich died. Knowing he was dying, Reich had avoided embarrassing his colleague, Nicolaus Copernicus. (Another component of the letters from Reich to Dantiscus is the references to either wine or beer, for which Canon Felix obviously had an appetite. It appears that one of the ploys that Dantiscus used to try to gain Reich's cooperation was the promise of libations.)

Unfortunately, Copernicus was not out of the woods. Two months later, on March 23, 1539, Dantiscus received a letter from another Warmia canon, Paul Plotkowski, concerning the women involved with his three wayward colleagues:

> As regards the Frombork wenches, Alexander's hid for a few days in his house. She promised that she would go away together with her son. Alexander returned from Lubawa with a joyous mien; what news he brought I know not. He remains in

> his *curia* with Niederhoff and with his *focaria* [mistress], who looks like a beer-waitress tainted with every evil. The woman of Dr. Nicholas did send her things ahead to Gdansk, but she herself stays on in Frombork . . .

As mentioned earlier, Niederhoff and Scultetus were the other two canons with mistresses, and Scultetus had a child with his partner, too. The judgmental Plotkowski was among the older canons at Frombork and had gotten off to a rocky start with Dantiscus. They had similar ambitions for promotion and had maneuvered against each other in recent years. Dantiscus had won. Still ambitious, and not stupid, Plotkowski was now trying to ingratiate himself with the bishop, and he knew that the "wenches" were something Dantiscus cared deeply about.

By the summer of 1539, Anna Schilling had become a priority for Dantiscus. He started seeking help from outside his own realm. He knew that Copernicus was very close to Tiedemann Giese, the bishop of the nearby district of Chelmno. Dantiscus sent Giese a letter in early July, having heard that Copernicus was on his way to visit him at his castle in Lubawa:

> He [Copernicus] is renowned and recognized far and wide, not only with distinction but also with admiration in many fields of fine writing. In his old age, almost at the end of his allotted time, he is still said to let his mistress in frequently in secret assignations.
>
> Your Reverence would perform a great act of piety if you warned the fellow privately and in the friendliest terms to stop this disgraceful behavior, and no longer let himself be led astray by Alexander

> [Scultetus], whom he declares to be all by himself outstanding in all respects among all our brothers, the officials and canons [of the Warmia Chapter].

Giese immediately wrote back to Dantiscus to say that he would convey his concern. Then two months later, he wrote to Dantiscus again:

> I have talked earnestly to Doctor Nicolaus about the subjects specified in your Reverence's warning. I put the situation, just as it is, before his eyes. He seemed to be disturbed not a little. For, although he has always obeyed your Reverence's wishes without delay, he is still falsely accused by malicious persons of secret assignations, etc. For he denies having seen her [Anna] since her dismissal, except that she spoke to him in passing as she was leaving for the fair in Królewiec. I was absolutely convinced that he is not as emotionally involved as most people think.

Giese perhaps protested too much when he stated, "I was absolutely convinced that he is not as emotionally involved as most people think." The canons at Frombork believed that Doctor Nicolaus was deeply attached to Anna.

COPERNICUS'S PLIGHT over Anna Schilling was genuinely serious, but it was certainly of his own doing. However, the second cause of Copernicus's intense stress in the winter and spring of 1539 was beyond his control. It was the great ferment of the first half of the sixteenth century—the Reformation.

9

The Taint of Heresy

Bishop Dantiscus's demand that Scultetus, Niederhoff, and Copernicus end their relationships with their mistresses bound the three canons together as if they were on collective probation, the black sheep of Frombork Chapter. In addition to Plotkowski, others eager to please the new bishop may have been making reports about them to Dantiscus. The sixty-six-year-old Nicolaus Copernicus, the nephew of perhaps the greatest leader Warmia had ever had, an active canon for almost four decades, a renowned doctor, and in some circles a famous astronomer, must have been appalled by all of the negative attention. Tiedemann Giese confirmed as much in his note to Dantiscus in July 1539, cited in the previous chapter: "I have talked earnestly to Doctor Nicolaus about the subjects specified in your Reverence's warning. I put the situation, just as it is, before his eyes. He seemed to be disturbed not a little."

The state of affairs was even worse than it appeared. In his letter to Giese, Dantiscus had referred to Copernicus's being "led astray by Alexander" Scultetus. By 1539 Scultetus was on a collision course with his superiors and those in his circle, including Copernicus, were at risk of being found guilty by association.

Scultetus must have been a charismatic individual, regarded as first among equals by some and a rabble-rouser by others within the group of canons at Frombork. Born Alexander Schultze in the city of Tczew, which was close to Gdansk and about fifty miles from Frombork, he, like Copernicus, had a doctorate in canon law. Scultetus was also a historian and cartographer (at one point, he and Copernicus began to create the first map of Prussia, and their work would form the basis of the first such map). Scultetus was younger than many of the other canons—his birth year is not known, but he died in 1564, so he was likely quite a bit younger than Copernicus. Scultetus was awarded the canonry left vacant when Andreas Copernicus died in 1518 (Dantiscus had also sought the post). Pursuing the position with great urgency, Alexander had circulated a formal letter of endorsement on which he gathered the signatures of thirteen supporters, one of whom was Andreas's brother. Nicolaus Copernicus enjoyed Scultetus's company and expressed as much to Dantiscus, perhaps during the fateful tour of 1538. It might have been a case of opposites attracting, because unlike Copernicus, who liked to stay in the background, Scultetus was bold, bordering on reckless.

He was certainly reckless in his contentious relationship with Bishop Dantiscus. Their personal animosity had started years before. As a young cleric and before his election to the Frombork canonry, Scultetus was stationed in Rome. For unknown reasons, while there he twice blocked attempts by Dantiscus to acquire a canonry in the 1510s. Then, when Dantiscus was finally awarded his Warmian canonry in 1529, Scultetus maneuvered to block any revenue that might flow to Dantiscus from the position, which was unheard of. Because of Scultetus's unwarranted scheme, the offended Bishop Ferber moved to have him excommunicated in 1531. However, the nimble Scultetus anticipated the move and secured formal protection from Rome, where he had many valuable connections from his earlier years. Not even an official letter of rebuke by Sigismund I, became bishop when Giese died in 1550. Hosius also later invited the Jesuits to enter Poland, which marked the beginning of the intense form of Catholicism characteristic of Poland ever since. His star shined during the Council of Trent in 1560, and in 1561 Hosius was made a cardinal.) Hosius would eventually be known as "the Hammer of the Heretics," "Death to Luther," and "the second Augustine." He knew that Scultetus had opposed his pursuit of the canonry, and just three days after moving to Frombork, wrote to Dantiscus, "With all my might, I shall fight and struggle to place his [Scultetus's] canonry in doubt rather than mine."

Several months after Hosius established himself in Frombork, Dantiscus and the king of Poland tried to secure a benefice for him. Through various machinations, Scultetus succeeded in blocking him from gaining it. In March 1539, Scultetus was elected cantor of the chapter, a largely honorary position, but one that carried additional income. However, a faction led by Hosius blocked him from taking the position. On April 14, Hosius himself was elected cantor. In May, the king sent an official letter to the Cardinal Protector of Poland in Rome, urging the pope to censure the meddling Scultetus. Then in early June, Dantiscus actually brought criminal charges against him—the charge was illegally engaging in business by selling flax.

IN THE MIDST of these aggressive and serious actions against Scultetus, the case against him took a significant turn. The first mention that he might be sympathetic to the Lutheran cause came from the future Hammer of the Heretics in the spring of 1539. In a note, Hosius said that Scultetus was "infected with the stain of Lutheranism." This would be the beginning of an

the king of Poland, could remove Scultetus once Rome h
ruled on the situation. Thus, there was essentially a stalemate
Scultetus and his Roman advocates aligned against Ferber, Dan
tiscus, and the king.

As already mentioned, Scultetus lived openly with a woman with whom he had at least one child. Yet when Dantiscus became bishop of Warmia, instead of laying low, Scultetus openly criticized him for obtaining additional benefices (essentially absentee positions that generated income). Scultetus's last attack on Dantiscus caused the bishop to write to Tiedemann Giese:

> I also hear that my benefice is being called into question as though I obtained it under false pretenses . . . In due course I shall see to it that those responsible for such mischief pay the penalty they deserve. I shall not lack the means suitable for this purpose . . . lest wicked men gain their pleasure, which can one day turn into a great grief for them.

Scultetus made a major blunder when he chose to make an enemy of Dantiscus. Instead of learning from his mistake, however, in 1538 he compounded it by opposing the new canon in Frombork, Stanislaw Hosius.

Hosius was the king's candidate for the vacant canonry, and he won the post in January of 1538. He moved to Frombork to commence his duties in July, just before Copernicus and Reich accompanied Dantiscus on the tour of the bishopric. Hosius was an impressive scholar and, like Dantiscus, a poet and writer. He was also a rabid anti-Lutheran, and within a few years would establish himself as the most active leader of the Catholic Counter-Reformation in Poland. (When Dantiscus died in 1548, Giese succeeded him as Bishop of Warmia, and then Hosius effort to brand Scultetus as a Lutheran and therefore a heretic.

The king received no response from the Cardinal Protector, so in May 1540, Sigismund took matters into his own hands. He ordered Scultetus to appear before him and answer to the charges that he was married, had children, and was a member of a heretical sect within the Catholic Church known as the Sacramentarians. This sect believed, as did the Lutherans, that the bread and wine did *not* become the body and blood of Christ during Mass. Scultetus disobeyed the order and did not present himself at the king's palace in Kraków. Because of this act of insubordination, the king banished him from the Kingdom of Poland.

Scultetus fled Warmia and took his family with him to Rome, where he believed he had strong support. In a hearing held there, he was acquitted of the charge of being a Sacramentarian. However, back home in Frombork, his enemies were still active. Scultetus's strongbox was discovered in his vacated house, and in late 1540 it was opened. What was found inside was indeed incriminating—two popular pieces of Reformation literature entitled "Commentary on Paul's Epistle to the Hebrews" and "Paul's Epistle to the Romans." Both of these works were written by the influential Swiss reformer Heinrich Bullinger, and Scultetus's copies contained many annotations in his own hand. This new evidence was sent to the king and then on to Rome. Even his protectors in Rome could not help him now, and in the winter of 1541, Scultetus was thrown in jail by the Roman authorities on the charge of heresy.

Thus, it appears that Scultetus was at least curious enough about Lutheranism to study its teachings, and he may have been sympathetic to the reformers. In these anxious times, it was easy to argue that he was guilty of something.

Being threatened with the label of Lutheranism was serious

business in Warmia. The bishopric was surrounded by Lutherans. Old-timers like Copernicus and Dantiscus must have been stunned by the swiftness with which the world they grew up in had changed. If Lutheranism swept over Warmia as it had other areas, the bishop and canons would lose everything—certainly their posts and possessions, and perhaps even their lives. From his correspondence, Dantiscus was clearly apprehensive about the threat to Catholicism all over Europe, but especially in his bishopric. He was also concerned about the risk of hostilities between the countries and kingdoms on either side of the religious line. On March 21, 1539, Dantiscus renewed Ferber's order from thirteen years earlier and outlawed all Protestants and their publications from Warmia. Then, in April 1540, the king of Poland declared Lutheranism illegal. Anyone found practicing it, or expressing public sympathy for it, was subject to prison and confiscation of property.

SHORTLY AFTER SCULTETUS departed Frombork with his family, his house was occupied by Paul Plotkowski's brother. This quick takeover was disputed by another Frombork resident, so the matter was brought to trial, with a committee of Frombork canons serving as the panel of judges. In a formal filing, Plotkowski, the brother of the canon, insisted that he would not go forward with the proceeding unless Copernicus and Niederhoff were removed from the presiding committee, since they were both under the same suspicion as Scultetus himself. This was an extraordinary charge, since it was made in a court of law.

What *was* Copernicus's attitude toward the Reformation? He left no written opinion, but an inference can be made. In 1523, a bishop from a diocese near Warmia published his own

110 theses, which were sympathetic to Martin Luther's positions. This action prompted Tiedemann Giese, Copernicus's devoted friend, to write a rebuttal. Giese's response was dedicated to Felix Reich and published in Kraków in 1525. It was a moderate document, preaching conciliation and acknowledging that the Lutherans were right about certain problems with the established church. Giese stated in the preface that his colleague Nicolaus Copernicus had urged him to have the document published. Most scholars therefore believe that Copernicus probably read and critiqued Giese's document prior to publication, and that it likely reflected his general attitude toward the reformers—he was sympathetic to their positions, but did not endorse breaking away from the established church.

Indeed, the position of the entire chapter of canons is uncertain. In the summer of 1540, Paul Plotkowski gave a speech before the assembled canons. He reminded them that in March of the previous year Bishop Dantiscus had issued his first edict outlawing the writings of the heretics and forbidding speeches that were favorable toward their positions. Then he reminded them of the second more emphatic anti-Lutheran proclamation issued just a couple of months previously. Plotkowski continued:

> His Reverence has also deemed it worthwhile to urge your Lordships to keep the royal edict before your eyes, to conform to the kingdom of Poland, his Reverence, and all the classes of these lands, and to publish a similar edict throughout the Chapter's domains. His Reverence cannot refrain from marveling that no such step has been taken up to the present time.

* * *

Before the events of 1538–39, it was difficult to imagine Nicolaus Copernicus reengaging with his manuscript and finding the motivation to finish and prepare it for publication. After all, even the generous letter from Cardinal Schönberg had not motivated Copernicus to do so. Now, given all the tension in Frombork and Warmia, and the level of distraction that Copernicus must have felt, a miracle would be needed if he was ever going to complete it.

10

The Catalyst

A few weeks after the August 1538 tour of Warmia, during which Dantiscus had first learned about Copernicus's mistress and ordered him to break off the relationship, Copernicus was back in Frombork trying to figure out what to do about Anna. One day during this anxious period for Copernicus, another mathematician sat three hundred miles away in the whitewashed, deliberately plain foyer of a spacious new house anticipating an uncomfortable conversation with *his* superior. This conversation would begin a chain of events that would change Western culture forever.

The other mathematician was Georg Joachim Rheticus. He was only twenty-four years old, yet already a mathematics professor at the University of Wittenberg, one of the premier universities in Europe and the heart of the Lutheran movement. He was waiting to meet with the head of the university, Philipp Melanchthon.

By September 1538, Melanchthon was acknowledged as the second most important individual behind the stunningly successful Lutheran Reformation, second only to the indefatigable Martin Luther himself. In fact, it is arguable that the Reformation would not have taken root if not for Melanchthon because he protected Luther from his most counterproductive impulses. Luther was boorish and incapable of tolerating different points of view, but Melanchthon was refined and understood that compromise was necessary to achieve broad support for their movement. His ability to bring opposing sides to the table had proved critical at several key moments in recent years. In 1530, Philipp drafted the Augsburg Confession, which remains the official creed of Lutheran doctrine. It secured Melanchthon's unique place in history.

The most famous portrait of Melanchthon, sketched by Albrecht Dürer in 1526, depicts a rather intimidating visage, with the reformer sporting a curly, somewhat unkempt beard and mustache, and a shock of longish hair that was just starting to recede, giving him a long forehead. He had a hawklike nose, high cheekbones, and fiery eyes underneath dark eyebrows (the actor Charlton Heston in the movie *Ben-Hur* resembles Melanchthon). But Melanchthon was a much softer soul than his appearance reflected. He was also a small man—an indifferent eater, Melanchthon was sickly skinny, almost skeletal. In an early letter, Luther mentioned that "we soon saw past his [Melanchthon's] shape and his appearance. . . . The only thing I am afraid of is that his frail constitution might not be able to endure the way of life in our region." Melanchthon looked very different from his friend and colleague Luther, who had always been heavy, and by 1538 was grossly fat.

Johannes Dantiscus knew and for a time admired both Luther and Melanchthon. In the early 1520s he had visited them in Wittenberg. He related:

> Luther rose, extended his hand to us and pointed at a seat. We sat down and we conversed about many things for more than four hours into the night. I found the man keen, learned, eloquent, but full of malice, arrogance and poisonous words against the Pope, emperor, and other princes . . . He has sharp

> eyes; they have something terrifying about them, as in obsessed people, like the King of Denmark, although I do not think that they were born under the same constellation. His language is vehement, spiced with coarseness and vulgarity.

The Reformation had made incredible progress by 1538, with many countries and principalities in northern Europe having officially broken away from the Catholic church and declared themselves Protestant. But this was still a very tense time for the Lutheran reformers. Just a few months before, the remaining Catholic princes of the German principalities had formed a Holy League to confront the Protestant League (the federation of Protestant principalities), and there was a growing fear that war among the various kingdoms was imminent. The savage Peasants' War (1524–25), fought only thirteen years earlier, was still a vivid memory, a repeat of which all antagonists wanted to avoid.

However, it was personal matters that had prompted Melanchthon to urge Rheticus to come to his home this September day. When Melanchthon greeted the waiting Rheticus, he most likely led him to the garden at the rear of the house. This was Melanchthon's favorite meeting place. The protective wall that surrounded the town of Wittenberg marked the end of the property. Beyond the wall flowed the gentle waters of the Elbe River, which meandered across the flat plain of Saxony in today's northeast Germany. The long, narrow enclosure was a peaceful place, filled with trees and plants that fascinated Melanchthon and on which he doted. He also kept an herb garden near the wall. There was a tiny grove to one side that contained a couple of benches and an odd granite table. It was here where Melanchthon would have directed Joachim to take a seat. Melanchthon informed Rheticus that as rector of the university, he was giving him, his youngest professor, a leave of absence from his teaching duties. He was to depart from Wittenberg immediately. Specifically, Melanchthon was sending Rheticus off to study with several of the most learned scholars in Europe. Philipp had made arrangements for Joachim to begin his leave in Nuremberg, where he would study with a close and talented friend, Johann Schöner. Schöner would then introduce him to others.

THAT RHETICUS had come so far at such a young age was remarkable given the cathartic event of his childhood. When he was fourteen years old, his father was executed in his hometown. Rheticus's father, a man named Georg Iserin, had been the doctor the previous fifteen years for the small town of Feldkirch, in the province of Voralburg, part of the Holy Roman Empire (Feldkirch is in western Austria today, just a few miles from the Swiss border; it is in the foothills of the Alps). He was a town leader, serving on several civic boards, and he had treated nearly every citizen at one time or another. Yet, something happened in 1528. The court records are obscure and confusing—he was condemned for one of two crimes: the practice of sorcery or theft. What is not obscure is what happened on the day the executioner came to Feldkirch. In Voralburg, the method of execution in the sixteenth century was beheading by sword. Rheticus's father was unceremoniously led up several steps to a raised wooden platform, and then forced to kneel. With no fanfare, the experienced executioner raised his sword and let fly a vicious swing, severing Iserin's head from his body. The teenage Rheticus probably did not actually witness the horror, but he certainly lived through the shock of his father's arrest and then the public humiliation that followed the execution.

Following his father's death, young Georg Joachim Iserin had to change his last name, an act required of the family members of executed prisoners. He eventually chose von Lauchen Rheticus, von Lauchen being a German version of his mother's maiden name, and Rheticus being a latinized form of the ancient name of the region that housed Feldkirch, Rhaetia. (By 1538, he had eliminated the "von Lauchen" and went by Georg Joachim Rheticus.) After the decapitation of his father, Rheticus, his mother, and his sister were allowed to remain in Feldkirch. It does not appear that they were prevented from taking part in the social conventions of the town. They were not ostracized, and they definitely had access to money.

Joachim himself was able to shake off the memories of that terrible day and return to his studies and thrive. While his father was still his tutor, he had shown a remarkable aptitude for mathematics, and he was able to develop this skill even further as he matured. Joachim attended the Latin school in Feldkirch for several years before his father's death. Afterward, his mother sent him to Zurich, about 100 miles away, and from 1528 until 1531, he was a student at a private school there. In Zurich, Rheticus began the intense study of Latin, Greek, the fine arts, and rhetoric; he was probably also exposed to the teachings of Ulrich Zwingli, the religious reformer. Rheticus later claimed to have met the mercurial, wandering scholar Paracelsus (Philippus Theophrastus), now acknowledged as one of the pioneers of modern medicine.

Joachim moved back to Feldkirch in 1531, and then fell under the influence of the new town doctor, Achilles Gasser. Among other intellectual pursuits, Gasser taught Rheticus the art of astrology and prognostication, as can be gleaned from a dedication that Melanchthon later wrote to Gasser: "Georg wanted me to dedicate this little book to you in particular so that he might show the memory of your old friendship, and that you are thanked through my voice, because you encouraged him to these studies, moved by the celestial significations . . ." Gasser's teachings were imprinted on the teenage Rheticus, because from this point forward, astrology would dominate his intellectual pursuits.

After just two years back in Feldkirch, and thanks to a letter of recommendation from Gasser (who was an alumnus), Rheticus was accepted as a student in the hotbed of the Reformation, the University of Wittenberg. For the next three years, from the fall of 1533 to the spring of 1536, Rheticus was a student there.

Wittenberg was a flourishing university community at this time. There were probably 3,000–4,000 inhabitants in the town itself, and about 800 students matriculated at the university. It gleamed with new buildings—the Lutheran movement was very good for the economy of Wittenberg and the surrounding area. The Castle Church, a gigantic structure joining the town's castle and cathedral, and the place where Luther posted his Ninety-five Theses in 1517, had been finished in 1509. The town's protective wall and moat were finished in the late 1520s, and the impressive Town Hall in the wide-open, cobblestoned Market Square was completed in 1535.

The layout of Wittenberg itself seemed to celebrate the university and the reformers. Built in a straight line, the town followed the course of the nearby Elbe River. The main avenue, Collegium-Strasse, connected the Castle Church to the town square. Further down the street, right beside one another, were the single large building of the university, Philipp Melanchthon's brand-new house (built by the Elector of Saxony himself), and Martin Luther's house, which was a huge converted monastery. Luther opened his door to many. A contemporary commented that "[a]n oddly mixed collection of young people, students, young girls, widows, old women and children live in the doctor's house, which is why there is great unrest in the house, for which reason many people feel sorry for Luther." The students rented rooms from the artisans and craftsmen who lived in the two- and three-story half-timber houses that lined the Wittenberg streets.

In addition to taking the usual courses in Greek, Latin, and rhetoric, Rheticus most likely took the math and science courses taught by Professors Johannes Volmer and Jacobus Milichius. Volmer was deeply immersed in the science of astrology, which may have been one of the reasons why Gasser urged Rheticus to matriculate there. That Joachim remained serious about astrology is certain, for his master's thesis was entitled "Do the Laws Condemn Astrological Prognostications?" In this disputation, presented on April 17, 1536, Rheticus argued that prognostications based on scientific astrology—the formal study of the stars—were not condemned by Roman law, though predictions based on superstitions were. Rheticus also argued that astrology should be used for major predictions only, such as the rise and fall of empires, not minor ones, such as when a farmer should begin to harvest his crops. Three years after entering the university, in the spring of 1536, Rheticus received his master's degree.

With only 800 students enrolled, even a busy man like Melanchthon was able to grow intimate with most of the students. And it is clear that Melanchthon had kept an eye on Rheticus, and that he was most impressed with his thesis on astrology. Upon graduation, Melanchthon startled Rheticus by offering him one of two new professorships in the sciences. Rheticus's science professor, Volmer, had died several months earlier, and the other science professor, Milichius, had resigned to study medicine. The other science position was given to Erasmus Reinhold, who was just three years older than Rheticus. He, too, would play a major role in sixteenth-century astronomy. Rheticus did not seek the appointment, and actually came close to declining the offer, feeling that he did not have enough mastery of the field. However, his many friends finally convinced him that he must take it.

So, by the time the students returned to classes in the fall of 1536, Joachim Rheticus, just twenty-two years old and himself a student only several months before, was the new Lecturer in Arithmetic and Geometry ("mathematicum inferiorum"). Melanchthon was not disappointed in his decision. Just as Rheticus had been an enthusiastic and impressive student, he was an equally impressive scholar. And though he was known to be more interested in research than teaching, Rheticus proved to be an inspiring instructor. One course in particular captivated Melanchthon—Rheticus's astronomy and astrology course. He used no textbook, preferring to distribute his own notes, which contained the horoscopes of thirty-two important individuals, including Luther and Melanchthon.

Rheticus's fundamental astrological beliefs were already in place when he delivered his master's thesis, and they can be easily distilled. First, he believed strongly in what was called natal astrology, or horoscopes. To cast proper horoscopes, it was essential to know the alignment of the planets and stars at the time of the person's birth. Rheticus realized that devising and reading horoscopes was not easy: "I concede however that the art is difficult, and that many erudite men have often been mistaken due to the diversities of the heavens, of the air, of the seeds, of the places, of education, of virtue, of custom, of life, etc." Second, he believed that certain individuals were chosen by God to be leaders and movers of history, and that the horoscopes of these individuals were therefore particularly important to understand. And third, he was not a strong proponent of "event" astrology, that is, the prediction of specific events, such as floods or wars. The following passage from one of Rheticus's later works essentially encapsulates his views:

> The first is the science of nativities which generally teaches from the figure of heaven which occurs in the hour of a person's birth, namely, what shall occur to him in his entire life and what will be said of him after he is dead. The other is the science of the revolutions of the years of the world and what occurs in the lower world in any given year, of good and bad, of heat and dryness. Yet among these two sciences, nativities is the more . . . useful. More useful because it is the duty of people to know what shall generally occur in all matters of this world. For which reason Ptolemy also says in the Almagest, "Stupid is he who is ignorant of his own nature."

Rheticus was an ideal science and mathematics professor for Philipp Melanchthon because he was talented, took basic science seriously, yet saw the main application of science precisely as Melanchthon did. Their shared vision was that the main role of science was to better understand God's message through nature, in particular, the heavens.

Melanchthon and Rheticus were hardly alone in their beliefs. The shock of the Reformation had shaken society to its very core, and the participants themselves almost could not believe what they were involved in. All of the early leaders of the Reformation were looking for affirmation that their revolution was predicted, either in scriptures or elsewhere, including the heavens.

However, Melanchthon was unusually enthusiastic. In fact, even Luther kidded him about his obsession with astrology. He once quipped that when Philipp talked about astrology he sounded like himself (Luther) after too many beers. Melanchthon peppered his lectures with comments about astrology, and made numerous references to it in his correspondence. An example of how it impacted his everyday thinking can be seen in a letter from 1535: "Sabinus [his son-in-law] is headstrong and will not listen to advice; this is due to the conjunction of Mars and Saturn at his nativity, a fact which I ought to have taken into account, when he asked the hand of my daughter." He once gave a speech entitled "On the Dignity of Astrology." During the speech, he argued that astrology was a true science, has great value, and that there is a need for greater understanding of the heavens in order to make better predictions. Melanchthon published a text in the physical sciences in 1549, and in that text he called for more research into the influence of planetary movement in human affairs.

For anyone who took astrology seriously, new astronomical ideas that allowed for a better understanding, and anticipation, of the movements of the planets would be very valuable indeed. Having someone like Rheticus around, one who could really analyze the heavens was important to Melanchthon. And if Rheticus's already considerable abilities could be improved, then efforts should be made to foster that improvement.

THUS, ONE REASON WHY Melanchthon was sending Rheticus away was to enhance his abilities as an astrologer. But, there was a second reason.

Though the University of Wittenberg was the wellspring of one of the most significant spiritual movements in history, it was still a university town, filled with young teenage boys barely past adolescence and largely unsupervised. Thus, it was a licentious environment. Ironically, Martin Luther's presence added to the anything-goes attitude because he was foul-mouthed and uncouth. In March 1538, the Elector of Saxony was forced to reprimand the university administrators:

> It has been reported to us that many improprieties take place and are to be found in our University of Wittenberg . . . But in spite of the fact that they [the faculty] have for that reason no particular industry, they rather let the youth do and go as they please and wish, through which their vice grows with their lack of teaching. And it is reported that an unindustrious student at times seduces another, which causes more grief and ignominy to the parents who send their sons and children there for discipline and teaching at their financial burden. When they again return home, they are made fun of and cursed, over which considerable unease arises among our citizens in Wittenberg, and has been felt for many years now. The youth also drastically overdress and are prodigal, and thereby burden their parents so much that it is talked of far and wide . . . all of which shall bring our university to destruction in a short time if it is not investigated.

Joachim Rheticus, as student and then young professor, was friendly with a group of young poets who were outstanding Greek and Latin scholars. Most likely, the connection was through several of the poets who were from Rheticus's region.

In early June 1538, a collection of epigrams was published in Wittenberg. The author was Simon Lemnius (1511–1550), who was from a town near Feldkirch. Lemnius was headstrong and cocky, and not overly impressed with the Reformation nor its leaders. The epigrams were meant to be barely veiled attacks on the main personalities of Wittenberg, including Luther and members of the faculty. The only ones who were not attacked were Melanchthon, who had befriended the poets, and the other poets, including Rheticus. The printing of the volumes was halted immediately by Luther, and the dozens that had come off press were ordered to be burned. Lemnius left town under cloak of darkness, just as he was ordered to stand trial. It was suspected that Melanchthon assisted his escape. Lemnius found refuge in one of the Catholic principalities.

When Lemnius continued his activities, Melanchthon fell into trouble for having supported him. Melanchthon withstood the immediate storm and would eventually gain Luther's complete forgiveness; Luther could not stay upset with Melanchthon for long, so deep was his admiration and affection. However, the same could not be said for Lemnius's friends, among them Rheticus. Melanchthon thought it prudent for Rheticus to leave Wittenberg until the Lemnius scandal settled. Fortunately, a productive exile presented itself.

On October 15, 1538, Melanchthon wrote to his dearest friend, Joachim Camerarius, to tell him about the impending visit from Rheticus:

> To the great Joachim Camerarius at the
> Academy of Tubingen, his best friend.
>
> Greetings! This youth is our professor of
> mathematics. He has a nature suitable to these arts
> and not abhorrent to the humanities. As he is above

all an astrologer, he has a strong command over that to which he is dedicated. Now he has gone forth to confer with Schöner and Apian on certain themes. Our mathematician wanted to greet you; he truly loves you greatly, not only due to your virtue and doctrine but also because of our friendship. I tell you this and beseech you, so that you will embrace him. He may not reach you for awhile, for which reason I write less . . . I also want you to compose an elegy against Lemnius. It should not contain insults but an honest and grave denouncement.

11

THE NUREMBERG CABAL

One morning in mid-September 1538, Joachim Rheticus and a traveling companion mounted their horses and waved good-bye to the friends who had gathered to see them off. The horses trotted down Collegium-Strasse, past the Castle Church, and then proceeded through Wittenberg's town gate.

Rheticus's companion was an undergraduate named Nicholas Gugler. He had been picked by Melanchthon to accompany the young professor and serve as his assistant for the duration of his leave. No one ever traveled alone in the sixteenth century, when brigands and highwaymen were a constant danger on the often-desolate roads, and the long hours of frequently numbing boredom demanded company. Melanchthon no doubt chose Gugler because he was both bright and a native of Nuremberg. He would be able to help Rheticus quickly familiarize himself with his new environment.

Once in Nuremberg, they were to present themselves, along with their letters of introduction from Melanchthon, to Johann Schöner, who was the headmaster of the secondary school there and one of the intellectual leaders of Nuremberg. Schöner would be Rheticus and Gugler's host.

Nuremberg was about two hundred miles southwest from Wittenberg. The travelers went by way of the main commercial road that originated in the northeast in Gdansk on the Baltic

Sea coast. With 20,000 inhabitants, Nuremberg was one of the largest cities in central Europe (the largest was Cologne, which had a population of 30,000; Nuremberg was about the same size as Ulm, Strassburg, Hamburg, Augsburg, and Lubeck). Nuremberg was home to the most distinguished collection of scholars, including astronomers, in the region. Not only was it densely populated, it was also wealthy, and it had an activist government determined to improve life for everyone within the city walls.

Rheticus and Gugler entered the woods and lands controlled by "the greatest, most famous, and best ordered of all the imperial cities" when they were still a day's ride away from the city itself, about twelve miles. As they drew closer, they passed one of the many man-made ponds and small lakes that were stocked with a wide variety of fish to make more tolerable the days on which meat was forbidden by religious custom. They probably saw one or two of the sandstone quarries that supplied most of the building stone used in the city as well. Then, when Rheticus and Gugler were about a half-mile away from Nuremberg, the woods came to an abrupt end and there was nothing but exposed ground the rest of the way. The open space was regulated and maintained by the city government for defensive purposes. Wars, rebellions, and uprisings were a constant threat in the sixteenth century, and the open ground made it impossible for an enemy to initiate a surprise attack.

Nuremberg was majestic. The city flowed down a steep hill, with the imposing Kaiserburg Castle at the top, and the entire town enclosed behind massive stone walls. Red-tiled roofs topped nearly every building. Rheticus, twenty-four years old, and Gugler, a teenager, entered Nuremberg through one of the city's six main gates after first crossing the 90-foot-long and 90-foot-deep moat by way of a movable oak plank bridge. The moat no longer contained water; it was filled with fruit trees and deer, and was used for games, contests, and strolls during peaceful times. When they emerged through the last of several nested gates, they faced a maze of more than 500 streets and thousands of houses and public structures. The streets were much broader than they had looked as the two Wittenbergers had approached.

They crossed the narrow Pegnitz River, which flowed right through the middle of town, over one of the half dozen or so bridges. On the river's banks they saw and heard the commotion caused by the tanners, butchers, and dyers who had their shops and homes by the water. The young men may also have noted the numerous public baths that drew their water from the river. And, of course, there were several mills that were powered by the current. Certainly, the main churches of St. Lorenz and St. Sebald could not have been ignored. The two men noted that the sandstone or half-timbered homes were adjoined and usually three or four stories high—the craftsman and artisans had their workshops on the first floor and their living quarters above. Climbing the hill, they went past the sprawling town-square marketplace, where generations earlier the Jewish Quarter had stood. It had been razed and cleared years ago, and the Jews completely expelled in 1498. The young men would have walked by a number of wells for drawing water (at least 100 in the city) and several of the distinctive fountains that the city leaders commissioned. And, assuming Rheticus and Gugler had timed it right, they might have seen the town's pigs being herded down to the river for their once-a-day watering. The city worked on the Great Clock system, in which the day was divided into day hours and night hours; thus, the first sixty-minute period after sunrise was the "one hour of the day." Bell ringers in the city marked the passing time.

After walking past St. Sebald and its dancing clock, and through the bustling marketplace, they would have arrived at the gymnasium where Johann Schöner was the headmaster.

In addition to being impressed by the size and efficiency of Nuremberg, Rheticus was keenly aware that he was in the town of the great Regiomontanus, the legendary astronomer and astrologer. A succession of astronomers in Nuremberg had committed themselves to maintaining and enriching Regiomantanus's legacy. As a result, Nuremberg retained its position as the most important central European city for astronomy and astrology.

Regiomontanus's first successor was Bernard Walther (1430–1504), Regiomontanus's wealthy benefactor, who had drawn him to Nuremberg in the first place. He became the literary executor of Regiomontanus's estate, and he continued the ambitious observational goals of the master. He made 746 solar observations and 615 planetary observations, a Herculean effort (by comparison, Copernicus recorded fewer than 100 observations in his career). Walther learned his lessons well and made careful comments about all aspects of the observation, noting the weather, cloudiness, humidity, time of day, and time of year.

Upon Walther's death in 1504, Johannes Werner (1468–1528) became the keeper of the Regiomontanus flame. Werner was a priest in a parish on the outskirts of Nuremberg proper and then within Nuremberg itself; Walther had been in his circle of friends. He controlled the Regiomontanus manuscripts after Walther's passing. Werner was a world-class mathematician who made fundamental contributions to trigonometry. He was also an outstanding cartographer and instrument maker.

Unfortunately, his astronomy was less rigorous. His book on the fixed stars prompted the letter from Copernicus to Wapowski mentioned earlier. This critical letter was surely read by the astronomer-astrologers in Nuremberg.

And now the legacy of Regiomontanus was protected in the person of Johann Schöner.

JOHANN SCHÖNER led a fascinating life, emblematic of these exciting, fluid, and also unsettled times. He was born in 1477, making him just four years younger than Copernicus. He attended the University of Erfurt, and eventually became a Catholic priest in 1515. While a priest in the town of Bamberg, Schöner built a print shop in his house, and he published books, made maps and globes, and produced astrological prognostications. He was a distracted and indifferent priest, and was eventually demoted and relocated to a lesser position in a smaller town. Then, when the Peasants' War broke out in 1524, he wasted little time in resigning from the priesthood—the peasants threatened to kill all priests, and many indeed perished before peace was restored. Schöner quickly converted to Lutheranism, and soon married. He knew Melanchthon through his astrology, and in 1526, Melanchthon appointed him headmaster of Nuremberg's new Gymnasium (Melanchthon was the founder of Germany's public school system, and he started gymnasiums, or secondary schools, throughout the German states). Schöner taught mathematics. He remained in that position until his death in 1547.

Schöner was an impressive polymath: he was a writer, editor, teacher, publisher, cartographer, globe maker, instrument maker, astrologer, and astronomer. That set of skills marked many of the astronomer-astrologers of this period. Schöner developed lasting fame for two innovative globes that he designed and built—one, from 1515, was among the first to show the New World, and it *was* the first to label it "America."

Deeply committed to astrology, Schöner was a rigorous, or scientific, astrologer. Rheticus would soon write: "When I was with you last year and watched your work and that of the other learned men in the improvement of the motions of Regiomontanus and his teacher Peurbach, I first began to understand what sort of task and how great a difficulty it was to recall this queen of mathematics, astronomy, to her palace, as she deserved, and to restore the boundaries of her kingdom." The "other learned men" referred to by Rheticus must have included several of Schöner's colleagues and friends: Peter Apian, Georg Hartmann, and Melanchthon's confidant, Joachim Camerarius. This passage shows that Rheticus, though already a professor of mathematics, arrived in Nuremberg as something of a neophyte in astronomy.

In addition to all his other talents, Schöner would go on to edit and publish thirteen of the books on the Index that Regiomontanus had drawn up during his lifetime. The most noteworthy was Regiomontanus's *Triangles of Every Kind* (1533).

SCHÖNER AND HIS NUREMBERG colleagues introduced Rheticus to a process that was to prove extraordinarily important—publishing. In the 1530s, Nuremberg was easily the center of book publishing in the German-speaking world, and vied with Venice and Paris as the most important publishing center in all of Europe. Rheticus had been exposed to publishing by his friend Achilles Gasser, and had seen its importance manifested in Wittenberg, where Luther and Melanchthon published everything from the seminal German Bible to textbooks, psalters, and broadsheets. But in Nuremberg, he was exposed to something else—scientific publishing.

In addition to publishing his own works, Schöner served as a scout for another publisher in Nuremberg, Johannes Petreius (1497–1550), the man who would be Copernicus's publisher. Rheticus got to know him well during his stay with Schöner. Yet another graduate of the University of Wittenberg, Petreius inherited a printing press shortly after finishing his studies and set up a printing and publishing business in Nuremberg in either 1522 or early 1523. It is not known why he chose Nuremberg, but the size of the city, its highly educated citizens, the fact that other publishers were making a comfortable living there, and the availability of skilled labor all played a role. By the middle of the 1530s he was acknowledged as the leading publisher-printer in the city.

Johann Gutenberg invented the movable-type printing press in the mid-1450s, and the earliest publishers were, in fact, printers. Unlike today's industry, in which printers manufacture books for publishers, there was no division of labor in the earliest stages of the industry. Printers sold the books that they produced directly from their shops.

Gutenberg's print shop had been in Mainz. Though he went bankrupt for his innovative efforts, his invention was recognized as genius right away. Within a few short years, there were printers all over Europe. One scholar estimates that by 1480, just twenty-five or so years after the invention, there were 110 printers in Europe, most in Italy and Germany. Petreius was among the first publishers to produce "difficult" books—books with illustrations and/or mathematics. He must have found them valuable works that sold at a good price because they became a prominent part of his list. He was also a very aggressive publisher, which was rare for the time. He actively scouted for "talent" and important manuscripts in the making, instead of sitting back and waiting for manuscripts to be brought to him. For example, when the mathematician, gambler, medical scholar, and astrologer Girolamo Cardano published the first of his mathematical works in 1539 with a publisher in Milan, he included a list of his planned or otherwise unpublished works, in an effort to find patrons. Petreius stepped forward and went on to publish many of Cardano's books, all of which were significant sellers.

What impresses one the most is the list of the other books and authors that Petreius published, which included works by Saint Augustine, Luther, Desiderius Erasmus (the most famous humanist in all of Europe), Ulrich Zwingli, Melanchthon, Henry VIII of England, Aesop, Joachim Camerarius, Regiomontanus, and others. In 1543, the year he published Copernicus's masterwork *On the Revolutions,* he also published Luther, Schöner, and Achilles Gasser.

THERE WAS ONE OTHER individual who was part of the circle of scholars who Rheticus befriended, and he was the most complicated of all. Andreas Osiander (1498–1552) played a key role in the success of the Lutheran Reformation in Nuremberg. Ordained a priest in 1520, Osiander converted to Lutheranism almost right away, and by 1522 was among the most strident Lutheran proselytizers in that city. His advocacy of the new religion helped to convince the city leaders to convert. As noted earlier, he was also instrumental in converting Grand Master Albert of Teutonic Prussia to Lutheranism, which happened in 1525.

Osiander was an intense man and was very vocal about his beliefs. Though he played a major role in winning over the city of Nuremberg—the first major population center to join with the reformers—over time he alienated Luther, Melanchthon, and all the other leaders of the movement; he also lost the faith of the city council of Nuremberg. By the late 1530s, when Rheticus was in town, he was searching for a new role. Fascinated by astrology and deeply involved with Petreius as well, he got to know Rheticus. He would later take a deep interest in the young professor's next intellectual adventure.

The months that Rheticus spent in and around Nuremberg in late 1538 and early 1539 were extraordinary. Living in Regiomontanus's city, visiting his observatory just outside the city walls, handling his manuscripts, and especially interacting with the men who were preserving his legacy—these stimulating days energized an already enthusiastic young man. Rheticus finished his study with a clutch of new mentors.

Someone, probably Schöner, had a copy of, or at least had read, Copernicus's short essay on the heliocentric theory penned all the way back before 1514. The scholars of Nuremberg had also read Copernicus's criticism of their late colleague Johannes Werner. Copernicus was clearly a topic of heated discussion because at some point Rheticus made the decision that he must go and meet this man:

> I heard of the fame of Master Nicolaus Copernicus in the northern lands, and although the University of Wittenberg had made me a Public Professor in those arts, nonetheless, I did not think that I should be content until I had learned something more through the instruction of that man. And I also say that I regret neither the financial expenses nor the long

journey nor the remaining hardships. Yet, it seems to me that there came a great reward for these troubles, namely, that I, a rather daring young man, compelled this venerable man to share his ideas sooner in this discipline with the whole world.

12

The Meeting

On May 14, 1539, Joachim Rheticus wrote to Johann Schöner in Nuremberg, telling him that he was presently in Posnan and on his way to try to meet Copernicus in Frombork. Several days later, he and his new traveling companion Henrich Zell found themselves on the Elblag-Braniewo road getting ever closer to Frombork. The horses galloped along a well-traveled dirt road that dipped and climbed on pleasantly undulating land with the Frisches Haff occasionally peeking through the trees and fields to their left. The flies following the horses, the birds in the leafy trees that lined both sides of the road, the bees hovering near the countless wildflowers, and the cacophony of noises made by the insects and animals in the woods made it clear to the travelers that spring had finally arrived in this northern region.

As they finally emerged from the woods into the open at Frombork, they could not fail to see the massive redbrick structure above them to the right. It completely dominated the small town that lay before them. The fifty-foot walls made the modest hill on which the structure was built appear clifflike. There was a narrow gate in the side of the fortress that they faced. Rheticus dismounted and walked up the little hill to enter the gate. When he found a servant, he was told how to find Copernicus.

One can imagine Rheticus's growing excitement after his

long journey. But he must have been puzzled, too. As intimidating as the fortress was, the town itself did not impress the visitor. The streets were narrow, the cottages small and nondescript, and the entire place smelled of fish. The smell came from the docks, which were several hundred yards from the road at the base of the hill. Rheticus was about to meet one of the legends of Europe—at least he was a legend among Rheticus's peers—a man who was perhaps the greatest astronomer and mathematician of his generation. What was he doing here in Frombork? If Copernicus was as talented as believed, then he had thrown away a life of acclaim and riches. A university town like Wittenberg provided forums and opportunities for intellectuals to interact with other scholars and grow. Large cities like Nuremberg were magnets for all kinds of bright and talented people who could assist a scholar with difficult puzzles. And, of course, Rome or any of the major capitals, such as Regiomontanus had enjoyed during his short life, were the pinnacle. Talented astronomers were in a position to become icons in such cities—making prognostications for kings, noblemen, and civic leaders; publishing books, calendars, and almanacs; and teaching courses. There were so many ways to make money and gain recognition. Rheticus had suspended his judgment until he saw Frombork, but now he was deeply confused. The next few weeks would do little to alleviate his confusion.

They continued down the road, with the town on their left and the fortress on their right. Now they could see the roof of the enormous cathedral behind the walls—it was as grand as any cathedral in any major city. Yet instead of sitting in a large town square in a city like Nuremberg or Toruń, or being attached to a castle as in Wittenberg, it was sitting here in tiny Frombork, in the middle of nowhere.

Toward the end of town, right near the long one-story building that Rheticus would soon learn was the hospital where Copernicus tended to his patients, a new road appeared to the right. The horses struggled as they climbed up the steep grade. Soon the south walls of the fortress appeared. There were several large houses scattered across the road from the fortress. As the road came to an end, Rheticus thought he knew which house must be Copernicus's. It was the last one, near the edge of the hill and almost looking down into the fortress.

Rheticus left his horse with Zell and walked across the grass field where goats and sheep were lounging in the midday sun, and he stood before the door. He and Zell were complete strangers in a town three hundred miles from their home in Wittenberg. Worse, they were technically in hostile territory, for only two months earlier the bishop of Warmia, Johannes Dantiscus, had evicted all Lutherans and closed the region to them; they were at risk of being thrown into prison. One year later, when Rheticus was still in Frombork, the ruling against Lutherans was compounded, this time by both the king and Dantiscus: "Under the penalty of losing head and property, of proscription or banishment from all royal lands, no one shall possess, read, or listen to the reading of Lutheran or similarly poisonous books, and all shall burn such books, booklets, songs, or whatever else has come from the poisonous places in the presence of the authorities."

The Lutheran Rheticus had defied the order—either out of bravery, ignorance, youthful hubris, or some combination of all three—and now stood at this spot unannounced and carrying only a satchel containing his clothes and effects, and a separate bundle of books. Rheticus's knocks on the wooden door were answered by one of Copernicus's servants. And finally, Rheticus met the man who he hoped held the secret to truly understanding the heavens.

Copernicus must initially have been wary of the young stranger at his door. Visitors, especially from distant regions, were rare in the small town of Frombork. Even rarer was a stranger seeking him, particularly one from the University of Wittenberg, who would have to be a Lutheran and have the favor of Luther and Melanchthon. Only a week or two earlier, Hosius had accused Scultetus of harboring Lutheran sympathies.

Rheticus most likely wasted little time before presenting Copernicus with the three bound books he had brought as gifts. Indeed, the volumes were very significant, and they began their relationship in a positive way. In the 1530s, books in general were expensive and valuable, and a gift of just one book was highly generous. A present of three books was extravagant. The books that Rheticus brought demonstrated that he was both a scholar and a mathematician, and that he surely understood Copernicus's pursuits, since these were precisely the kind of books that Copernicus appreciated and perhaps even needed. We know that Copernicus did not have these particular books and did use them, at least one extensively, during the remaining years of his life. Rheticus had chosen well.

The first bound book contained two works, a 1533 Greek edition of Euclid's *Elements* coupled with Regiomontanus's *On Triangles of Every Kind*, edited by Johann Schöner. The Euclid volume had been published in Basel in 1533, and the Regiomontanus volume was published by Petreius, also in 1533. These two combined volumes represented the most important work on geometry, coupled with the most important work on trigonometry. The second bound book contained three works, Peter Apian's *Instrumentum primi mobilis*, which was published by Petreius in 1534, Geber's *De Astronomia Libri IX*, published by Petreius in 1534, and Witelo's *Perspectiva*, also published by Petreius (1534–35). The third and final book was a major prize, Ptolemy's *Almagest*, specifically the Greek edition published in 1538 (the Greek title was *Syntaxis*; it was published in Basel). Rheticus had inscribed all three volumes, indicating that they were gifts. (It appears, based on another inscription, that the copy of the *Almagest* was previously owned by one of Rheticus's professors at Wittenberg, Jacobus Milichius.)

Copernicus already owned a copy of Euclid's *Elements*, but his volume was a Latin translation from his undergraduate days (published in 1482); the newer version, in the original Greek, would have been most welcome. Regiomontanus's *On Triangles of Every Kind* was a treasure, for the canon did not have a copy, though he did own a copy of Regiomontanus's *Table of Directions*, published in 1490. In addition, the Greek translation of Ptolemy would have made an immediate impact on the astronomer. He already possessed a Latin translation, based on Gerard of Cremona's thirteenth-century work (published in 1515), but the official Greek edition would have been coveted for checking certain assumptions. It is very curious, by the way, that Copernicus acquired a copy of Gerard of Cremona translation, but not a copy of the 1496 edition of Regiomontanus/Peurbach's *Epitome of the Almagest* (though he was of course very familiar with the *Epitome*).

Copernicus's personal library was remarkably small, and Rheticus's gifts added substantially to it. This is yet another mystery surrounding Copernicus. It is known that he had access to important volumes owned by his colleagues and friends, but Copernicus had the money to buy his own books, yet chose not to. The books that he did own he made copious notes in, as was common among scholars in the sixteenth century. But he could not make notes in the borrowed books.

It appears that Copernicus quickly warmed to his young visitor. He probably showed him around town, the cathedral grounds, his tower, and his observation area during the first days after his arrival.

Frombork was physically about the same size as Rheticus's hometown of Feldkirch and new hometown, Wittenberg, but with only about a thousand inhabitants, it was much less densely populated. Though Feldkirch had the hills and Alps, and Wittenberg had the Elbe, Frombork had the Frisches Haff, the sandbar, and the Baltic Sea beyond. It probably did not take Rheticus long to realize that Cathedral Hill (and the canons attached to it) cast a literal and figurative shadow over everything that transpired in the small village.

As a fellow astronomer, Rheticus would have realized that the catwalk on the ramparts along the cathedral wall was not wide enough for astronomical instruments. Even more problematic was the wind, which would have made careful use of delicate instruments impossible. The cathedral courtyard was also unworkable for observations because of the high walls surrounding it. Instead, Copernicus had built a simple patio near his house more than twenty-five years earlier, and this was where he formally observed the heavens. In the winter of 1513, Copernicus had bought 800 bricks and a barrel of chlorinated lime from the cathedral's brickyard and lime house, likely the raw materials for the patio.

Copernicus did his groundbreaking work with just a few primitive instruments; Rheticus had just left Nuremberg, where he had befriended the finest astronomical instrument makers of the age. Copernicus's primary workhorse instrument was a triquetrum. It consisted of three wooden components—an upright pole about twelve feet tall; a second piece, hinged to the pole, that had two sights on it (as on a gun barrel) and was used to find the planet or star to be observed; and a third piece that measured angles. Copernicus called the triquetrum the "Ruler of Ptolemy." Years later, one of Copernicus's greatest admirers, Tycho Brahe, sent his assistant to Frombork to purchase the canon's triquetrum so that he could better understand how Copernicus went about his observations. Back in Denmark and working with his hero's tool, Brahe was incredulous. He wrote that the eyepiece was so imprecise that "an error of several minutes can occur. Hence it is a wonder how not only Copernicus but also the ancients, who used such eyepieces, could have attained any precision, even if everything else was in perfect order."

A second instrument Copernicus used was a quadrant, which he later described at great length in *On the Revolutions*. A quadrant looked like a sun dial mounted on a wall; its vertical base was quite large, with a radius of "five or six square feet." This large instrument was used to note the summer and winter solstices.

Copernicus's third instrument was a spherical astrolabe, which consisted of nested circles made of brass. The astrolabe was used to chart heavenly bodies by determining their celestial longitude and latitude.

Rheticus had learned about the *Commentariolus* in Nuremberg, and in that document, as we know, Copernicus alluded to his larger work that would provide the proof of the startling assertions in the short treatise. Rheticus most likely asked the aging canon about it soon after arriving in Frombork. Copernicus must have been eager to show the work of countless hours to a mathematician capable of understanding its significance. This may have been the first time that a talented mathematician examined the handwritten pages filled with mathematics, dozens of geometric drawings, and tables of data.

(It is probable that Bernard Wapowski read parts of the manuscript when he obtained from Copernicus the planetary tables mentioned in Chapter Six, but Wapowski was not the mathematician that Rheticus was.) It is certain from everything that followed that Rheticus was immediately impressed. He now knew that the legend surrounding the canon was accurate. He would later write, "This man whose work I am now treating is in every field of knowledge and in mastery of astronomy not inferior to Regiomontanus. I rather compare him with Ptolemy . . ."

13

The First Summer

By comparing Copernicus to Regiomontanus and Ptolemy, Rheticus conveyed that he was greatly impressed with Copernicus's masterwork. In the pages of Copernicus's manuscript, Rheticus observed the manifestation of a brilliant mind that was also patient, diligent, and respectful of the great astronomers who had preceded him. It was a long work—the English translation fills nearly 400 book pages—composed of lengthy tables with data on the motions of the heavenly bodies, numerous geometric diagrams, page after page of trigonometric and algebraic expressions, instructions on how to build and use astronomical instruments, and, of course, a description of his new model of the heavens. Rheticus gushed: "But from the time that I became, by God's will, a spectator and witness of the labors which my teacher performs with energetic mind and has in large measure already accomplished, I realized that I had not dreamed of even the shadow of so great a burden of work. And it is so great a labor that it is not any hero who can endure it and finally complete it."

The manuscript was divided into six parts. The first was a "general description of the universe." The outline of the sun-centered universe was presented in this early section. The second explored the "doctrine of the first motion," a term used to describe the perceived motion of the fixed stars, the specks of light in the night sky that always retained their position in relation to other heavenly bodies. The third book focused on the the sun; the fourth addressed the moon and lunar eclipses; the fifth, the motions of the other planets; and the sixth discussed the celestial latitudes of the planets.

From everything that transpired over the next several months, two developments must have occurred. First, Rheticus quickly realized that the manuscript that he was reading was extraordinary in its originality and ambition, and he informed Copernicus of his judgment. Second, Rheticus asked Copernicus if he could write a manuscript that attempted to summarize and explain what was in several major sections of the work. Copernicus agreed. Over the next sixteen weeks, from late May to late September, Rheticus read and absorbed most of the complicated manuscript, conferred with Copernicus often, and then wrote a refined and polished manuscript containing his own summary of Copernicus's system. The resulting book, entitled the *Narratio prima* (the First Report), was to introduce to the world the canon's heliocentric theory. (I have converted most book titles cited in this book from their original Latin to an English equivalent, but I will keep referring to Rheticus's book as the *Narratio prima*.)

In the pages of the *Narratio prima* and in a preface to a book that he wrote a little over a decade later, Rheticus provided several fascinating glimpses of how Copernicus worked, how the two astronomers interacted with each other, what they talked about during these watershed months, how Rheticus himself worked, and what he did during his months in Warmia. The portrait of Copernicus and the buoyant tone of the *Narratio* show that both men were enthralled to be engaged with each other.

Rheticus's reaction to Copernicus's manuscript energized the aging astronomer. Whereas every other description of Copernicus's personality depicts him as dour, Rheticus stated that "my teacher [Copernicus] was social by nature." When Rheticus told Copernicus that he had been invited to visit the mayor of Gdansk, "he [Copernicus] rejoiced for my sake and drew such a picture of the man that I realized I was being invited by Homer's Achilles, as it were." Nowhere else in all the documents dealing with Copernicus does he ever appear this effusive. One cannot rule out the possibility that Rheticus's descriptions may have misrepresented certain interactions, but the success of their collaboration suggests not.

Outside Copernicus's house the atmosphere was turbulent and inhospitable for both men. By remaining in Frombork, the Lutheran Rheticus was flouting Bishop Dantiscus's recent edict expelling all followers of Luther. He could be forced from Warmia at any moment. Copernicus's situation was even more uncomfortable. He was still being watched by his fellow canons and townspeople because of the Anna Schilling controversy. Anna had probably vacated the canon's house by May 1539, but she might still have been living in Frombork. Worse, Alexander Scultetus was feeling the full fury of Dantiscus's and Hosius's rage. The very month of Rheticus's arrival was when Dantiscus had the king of Poland send a letter to Rome urging Scultetus's censure, and Hosius for the first time wrote of Scultetus's possible sympathy with Lutheranism.

Yet, inside Copernicus's house, the two men were able to ignore these distractions and focus on what had brought them together—their mutual love of astronomy. Rheticus had just left the intellectually energized atmosphere of Nuremberg, so he was continuing his immersion in astronomy. But for Copernicus, collaborating with Rheticus was the first time he'd worked directly with a gifted mathematician since he had worked with Novara four decades earlier. Copernicus showed

Rheticus how he went about synthesizing his results with those of the astronomers that had come before him:

> I see that my teacher always has before his eyes the observations of all ages together with his own, assembled in order as in catalogues; then when some conclusion must be drawn or contribution made to the science and its principles, he proceeds from the earliest observations to his own, seeking the mutual relationship which harmonizes them all; the results thus obtained by correct inference under the guidance of Urania he then compares with the hypotheses of Ptolemy and the ancients; and having made a most careful examination of these hypotheses, he finds that astronomic proof requires their rejection; he assumes new hypotheses, not indeed without divine inspiration and the favor of the gods; by applying mathematics, he geometrically establishes the conclusions which can be drawn from them by correct inference; he then harmonizes the ancient observations and his own with the hypotheses which he has adopted; and after performing all these operations he finally writes down the laws of astronomy.

Rheticus left a description of the two men working together, the young man being more interested in the results and astrological implications, and the older and wiser one urging patience and the importance of the fundamental scientific principles:

> I remember myself being driven by juvenile curiosity. I wished to hasten to the stars' sanctuary.

> So, agreeing with the very good and very great Man, Copernicus, I sometimes blamed the painstaking attention to details. But he was bewildered by my soul's honest thirst, and with a soft arm, he used to exhort me to take my hand off the Tables. "Personally," he said, "if I could get the truth from a sixth part which is an increment of 10 minutes, my spirit would exult as much as when was received the discovery of the formula of the ratios by Pythagoras."

Rheticus continued:

> He wanted his researches to be above the average. That is why he avoided grindings, not by inertia nor by fear of boredom. Some people seek and even require those little grains, like Peurbach in the subtlety of his tables of eclipses. They can see in them all the care taken to locate the stars with precision. While they are impressed by the seconds, thirds, fourths, fifths, and little divisions, they forget the integer parts, not giving them a single look. And in the small interval of times of the "Phenomena," they often are wrong by hours, and sometimes, entire days. There is an amazing fable of Aesop where an order is given to search for a lost cow. It is found, but the men who are to bring it back, see little birds and go after them, forgetting the cow.

Rheticus's reference to the famous Aesop fable is another way of saying that other astronomers were guilty of focusing on the individual components and missing the big picture. Not Copernicus.

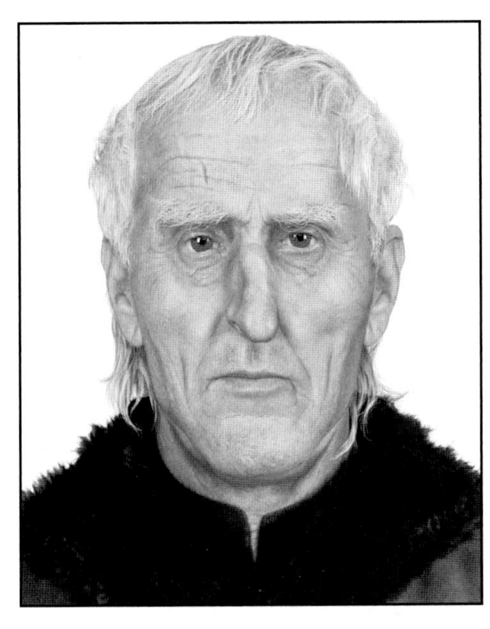

In 2005, researchers found what they believe to be the skull and skeleton of Copernicus beneath the floor of the cathedral in Frombork. This image of Copernicus as he looked at the age of seventy is based on a facial reconstruction by forensic experts.
(Corbis)

The Teutonic Knights slaughtered the native Prussians in the Baltic region in the thirteenth century and then founded the cities and towns where Copernicus lived in the fifteenth and sixteenth centuries.
(Corbis)

The façade of Copernicus's first childhood home as it looks today, in Toruń in present-day Poland. (PHOTOGRAPH BY JACK REPCHECK [JR])

One of the medieval gates in the wall surrounding Toruń. (JR)

The bishop of Warmia's castle in the town of Lidzbark. Copernicus lived here with his benefactor and uncle, Bishop Lucas Watzenrode, 1503–1510.
(JR)

Lucas Watzenrode, the bishop of Warmia, 1489–1512.
(Courtesy of the Nicolaus Copernicus Museum, Frombork)

Frombork, the village where Copernicus lived for most of his adult life. The villagers worshipped in the church of St. Nicolaus. In the background is the Frisches Haff (today called Vistula Bay), a freshwater lagoon that opens onto the Baltic Sea.
(JR)

Cathedral Hill, where Copernicus and his fellow canons lived and ruled the surrounding countryside.
(JR)

Copernicus's tower, a three-story structure built into the defensive walls of the fortress; this is where Copernicus lived when he was within the Cathedral Hill complex. (JR)

The site of Copernicus's house. The current house sits on the foundation of the one in which Copernicus resided just outside the walls of the fortress. The great astronomer probably made many of his observations from this location. (JR)

Copernicus's favorite astronomical instrument—the triquetrum, or "ruler of Ptolemy"—which he used for many observations. (JR AND COURTESY OF THE NICOLAUS COPERNICUS MUSEUM, FROMBORK)

Regiomontanus, born Johannes Müller, was the greatest astronomer of the fifteenth century. Copernicus studied several of his important books. (CORBIS)

Philipp Melanchthon, one of the catalysts of the Lutheran Reformation, who gave Joachim Rheticus his leave of absence from the University of Wittenberg. This allowed Rheticus to find and work with Copernicus in Frombork. (CORBIS)

Tiedemann Giese, Copernicus's fellow canon and best friend. For many years, Giese urged Copernicus to publish his *On the Revolutions of the Heavenly Spheres*, but only Rheticus finally persuaded him. (COURTESY OF THE NICOLAUS COPERNICUS MUSEUM, FROMBORK)

Johannes Dantiscus, bishop of Warmia 1538–48. Dantiscus made Copernicus's last years very uncomfortable. (COURTESY OF THE NICOLAUS COPERNICUS MUSEUM, FROMBORK)

NICOLAI CO-
PERNICI TORINENSIS
DE REVOLVTIONIBVS ORBI-
um cœlestium, Libri VI.

Habes in hoc opere iam recens nato, & ædito, studiose lector, Motus stellarum, tam fixarum, quàm erraticarum, cum ex ueteribus, tum etiam ex recentibus obseruationibus restitutos: & nouis insuper ac admirabilibus hypothesibus ornatos. Habes etiam Tabulas expeditissimas, ex quibus eosdem ad quoduis tempus quàm facillime calculare poteris. Igitur eme, lege, fruere.

Ἀγεωμέτρητος οὐδεὶς εἰσίτω.

Norimbergæ apud Ioh. Petreium,
Anno M. D. XLIII.

The title page of one of the most influential books ever published, *On the Revolutions of the Heavenly Spheres,* which appeared in the spring of 1543. Copernicus died on the very day that he received his first copy. (CORBIS)

* * *

WORD WAS RELAYED to Tiedemann Giese, Copernicus's best friend, that a young Lutheran mathematician had traveled hundreds of miles to learn about the doctor's astronomical work. Giese probably also heard that tongues were wagging in Frombork over yet another possible indiscretion by the canon. When Giese subsequently learned that Rheticus had fallen ill, he promptly urged the two men to visit him at his castle in Lubawa, ostensibly for Rheticus to receive the fresh air that he needed, but also to get the two men out of Frombork—"I had a slight illness and, on the honorable invitation of the Most Reverend Tiedemann Giese, bishop of Chelmno, I went with my teacher to Lubawa and there rested from my studies for several weeks." Copernicus had stayed at the castle just two months earlier, in late April and early May, when he traveled to treat Giese for an illness. The two astronomers arrived in Lubawa in early to mid July.

Though Giese succeeded in giving Copernicus some relief from the tension in Frombork, he could not keep all of it away. In a letter cited earlier, dated July 5, 1539, Dantiscus mentioned to Giese that he heard that Copernicus was on his way to visit him: "I have heard that your Reverend will receive the distinguished and very learned man, Doctor Nicolaus Copernicus, whom I truly cherish not otherwise than as a brother." Dantiscus continued:

> In his old age he is still said to let his mistress in frequently in secret assignations. Your reverence would perform a great act of piety if you warned the fellow privately and in the friendliest terms to stop this disgraceful behavior, and no longer let himself

> be led astray by Alexander [Scultetus], whom he declares all by himself outstanding in all respects among all our brothers, the officials and canons.

Bishop Giese dutifully had a discussion with Copernicus about Anna and Alexander and conveyed that news back to Dantiscus. Copernicus had to respond at least one more time that summer to the issue that would not go away. His fellow canon Achacy von Trenck visited Lubawa while Copernicus and Rheticus were there and he reported to Dantiscus on September 13, 1539, that "[w]hen Doctor Nicolaus, whom I found at Lubawa, heard his housekeeper mentioned, he declared that he would never receive her in his house nor do anything further in this case. I know that he was advised by the Reverend Bishop of Chelmno to do so, I hope not in vain, since his age and prudence can readily deter an upright, good man from actions of this kind in the future."

Even these distractions were not going to impair this fruitful period. Giese had also urged the two men to journey to Lubawa so that the bishop could revisit the topic that Copernicus had waved off many times in the past: the publication of his astronomical manuscript. With Rheticus there to support him, perhaps his plea would be heard. In probably the single most meaningful scientific exchange found in the Copernicus sources, Rheticus recorded a discussion between Copernicus and Giese about how to publish his important work. It is difficult to discern whether Rheticus witnessed a single discussion, or whether his account was a conflation of years' worth of prodding by Giese. Chances are that Rheticus observed at least one discussion in which all the issues were aired while with Giese at Lubawa, but it was surely not the first time that such a discussion had taken place.

Rheticus even suggests that Giese may have been responsible for Copernicus's taking up astronomy seriously again when he moved to Frombork in 1510. In the *Narratio,* Rheticus says:

> He [Giese] realized that it would be of no small importance to the glory of Christ if there existed a proper calendar of events in the Church and a correct theory and explanation of the motions. He did not cease urging my teacher, whose accomplishments and insight he had known for many years, to take up the problem, until he persuaded him to do so. Since my teacher . . . saw that the scientific world also stood in need of an improvement of the motions, he readily yielded to the entreaties of his friend. He promised that he would draw up astronomical tables with new rules and that if his work had any value he would not keep it from the world.

Rheticus goes on to say that Copernicus had long known that the prevailing theories of the heavens were in need of revision, and that the ideas needed to correct them would, in a great understatement, "contradict our senses."

In past discussions, Copernicus had stated that he was content to leave the manuscript unpublished, and to make available only the tables with no proofs of the rules implicit in the tables. Copernicus argued that this way "philosophers" (i.e., astrologers) and "common mathematicians" would have the basic information about planetary positions that they needed for their work, but because the underlying hypothesis—that the sun is the center of the universe—would be unstated, there would be no controversy. Copernicus then argued that "true scholars" would know in an instant ("would easily arrive" were his actual words) the underlying thesis, but it would remain a secret among this small number of gifted mathematicians "upon whom Jupiter had looked with unusually favorable eyes." Thus, "the Pythagorean principle would be observed that philosophy must be pursued in such a way that its inner secrets are reserved for learned men, trained in mathematics."

Copernicus argued that scholars had deduced the underlying principles to the *Alfonsine Tables.* Giese pointedly disagreed and said that the tables by themselves had led to confusion and errors. To say " 'The Master says so'—a principle which has absolutely no place in mathematics," is a lazy way out, he argued. Even Aristotle ultimately had to rely on mathematicians to prove his contentions about the movements in the heavens. But Giese was also realistic and told Copernicus that though he must publish the proofs, he must also be prepared to be rebuffed by other astronomers clinging to their own beliefs. Rheticus concludes: "By these and other contentions, so I learned from friends familiar with the entire affair, the learned prelate won from my teacher a promise to permit scholars and posterity to pass judgment on his work. For this reason men of good will and students of mathematics will be deeply grateful with me to His Reverence, the bishop of Chelmno, for presenting this achievement to the world." Rheticus was being generous to Giese, for it would be the young Lutheran professor who would convince Copernicus to present his "achievement to the world."

14

Convincing Copernicus

TOWARD THE END OF THE SUMMER, Copernicus and Rheticus left Giese in Lubawa and returned to Frombork. There Rheticus put the final touches on his manuscript about Copernicus's theory of the heavens, the closing words being, "From my library at Frombork, September 23, 1539." It is not known where Rheticus's library was, but it was most likely a room in the canon's house. After finishing the manuscript, he and Heinrich Zell traveled to Gdansk, the largest city in the area and probably the only one with a printer. Rheticus gave the stack of paper to Franciscus Rhodes to print and publish. It appears that Zell stayed in Gdansk to supervise the process and proofread the pages.

Rhodes must have been an efficient printer because the book was ready for sale by April of 1540. Thus, it took only eleven months from the moment when Rheticus knocked on Copernicus's door on that memorable day the previous May until there was at last a description of the science of the Canon of Frombork. Those intellectuals who had been waiting for it were not disappointed—the *Narratio prima* was a splendid work. Though written quickly and then published without delay, it was nonetheless smoothly composed and riveting to read. To this day, it is arguably the best primer on the heliocentric theory of Copernicus. The *Narratio* was doubly impressive because it was not a straight summary of the canon's larger work. Instead, Rheticus was able to absorb the wealth of detailed and technical information in Copernicus's manuscript and then construct his own narrative in order to address the skepticism that he knew would greet the theory.

The short book (in English it is about 100 pages long, significantly longer than the *Commentariolus*) was written in Latin as a long and formal letter addressed to "The Illustrious Johann Schöner, as to his [Rheticus's] own revered father." Though Schöner may not have urged the young Lutheran to travel all the way to Frombork to seek Copernicus, it was Schöner's fascination with the canon that had planted the seed in Rheticus's mind. Rheticus continues: "I promised to declare as early as I could, whether the actuality answered to report and to my own expectations . . . To fulfill my promises at last and gratify your desires, I shall set forth . . . the opinions of my teacher." Rheticus does not refer to Copernicus by name; he is called "my teacher" throughout (though the canon's name was given on the book's title page).

The Wittenberg professor early on gives Copernicus the highest compliment that could be bestowed on an astronomer, and thus prepares the reader for the significance of what is to follow:

> First of all, I wish you to be convinced, most learned Schöner, that this man whose work I am now treating is in every field of knowledge and in mastery of astronomy not inferior to Regiomontanus. I rather compare him with Ptolemy . . . My teacher has written a work of six books in which, in imitation of Ptolemy, he has embraced the whole of astronomy, stating and proving individual propositions mathematically and by the geometric method.

Rheticus is careful throughout his book to compliment Ptolemy, Aristotle, and all the ancients. He never describes Copernicus as a revolutionary—rather Rheticus shows him as a respectful astronomer who assiduously studied every available source and was committed to the evidence. He also makes many references to God as the divine creator of the heavens, to the Roman gods, to Aesop's fables, and to Regiomontanus. Although there are several passages about astrology, the zodiac, and horoscopes, the focus of the book is on Copernicus's theory of the movements of the heavenly bodies.

The *Narratio prima* is a summary of Copernicus's manuscript, but Rheticus does not discuss heliocentrism and the moving earth right away. He begins with a brief on the motion of the fixed stars, stressing the importance of the observations Copernicus made in Italy and also one he made in 1525. Next he discusses the tropical year, which was critical information for calendar reform. Toward the end of this section, Rheticus gives the first hint of what is to come: "That there necessarily was a deficiency of 19/20 of a day from Hipparchus to Ptolemy, and from Ptolemy to Albategius of about 7 days, I have deduced, not without the greatest pleasure, most learned Schöner, from the foregoing theory of the motions of the stars and from my teacher's treatment of the motion of the sun, as you will see a little further on."

Then, a dozen pages into the *Narratio*, Rheticus displays his astrological side.

> I shall add a prediction. We see that all kingdoms have had their beginnings when the center of the eccentric was at some special point on the small circle. Thus, when the eccentricity of the sun was at its maximum, the Roman government became a

> monarchy; as the eccentricity decreased, Rome too declined, as though aging, and then fell. When the eccentricity reached the boundary and quadrant of mean value, the Mohammedan faith was established; another great empire came into being and increased very rapidly, like the change in the eccentricity. A hundred years hence, when the eccentricity will be at its minimum, this empire too will complete its period. In our time it is at its pinnacle from which equally swiftly, God willing, it will fall with a mighty crash. We look forward to the coming of our lord Jesus Christ when the center of the eccentric reaches the other boundary of mean value, for it was in that position at the creation of the world. This calculation does not differ much from the saying of Elijah, who prophesied under divine inspiration that the world would endure only 6,000 years, during which time nearly two revolutions are completed. Thus it appears that this small circle is in very truth the Wheel of Fortune, by whose turning the kingdoms of the world have their beginnings and vicissitudes. For in this manner are the most significant changes in the entire history of the world revealed, as though inscribed upon this circle.

Curiously, just one year after the *Narratio* appeared, Martin Luther prepared a chronology of the world in which he predicted the end of the earth after 6,000 years and cited the same prophesy by Elijah.

About one-quarter of the way into the short book, Rheticus finally clears his throat and begins to discuss the highlight of Copernicus's work. He reluctantly states that Ptolemy's model cannot explain the movements of celestial bodies in a consistent way, so "[i]t was therefore necessary for my teacher to devise new hypotheses . . ." A few pages later, the heliocentric theory is revealed for the first time in a published book:

> The planets are each year observed as direct, stationary, retrograde, near to and remote from the earth, etc. These phenomena, besides being ascribed to the planets, can be explained, as my teacher shows, by a regular motion of the spherical earth; that is by having the sun occupy the center of the universe, while the earth revolves instead of the sun on the eccentric, which it has pleased him to name the great circle. Indeed, there is something divine in the circumstance that a sure understanding of celestial phenomena must depend on the regular and uniform motion of the terrestrial globe alone.

Rheticus goes on to discuss the second stunner in Copernicus's model—"the earth, like a ball on a lathe, rotates from west to east, as God's will ordains; and that by this motion, the terrestrial globe produces day and night . . ."

IN MID-FEBRUARY 1540, a batch of preliminary pages from the *Narratio* was sent to Philipp Melanchthon in Wittenberg. No doubt Rheticus wanted to inform the University of Wittenberg's rector about his activities, and he also knew that the several references to astrology would likely please Melanchthon. Another set of advance pages was sent to Andreas Osiander in Nuremberg.

Osiander wrote back to Rheticus immediately. He was

clearly captivated by what he had read. After a lengthy comment on Rheticus's passage in his book about the second coming of Christ and the determining of the precise age of the earth, he says, "Yet enough of this, of that above, just as I ask you again and again that you offer me your friendship, I ask that you apply yourself diligently and win over the friendship of that man [Copernicus] for me as well. I don't risk writing him at the present, and although I didn't intend to, you will certainly keep these my triflings from him."

Rheticus and Giese were determined to use the *Narratio* as advance publicity for their mentor and friend. Immediately after obtaining their first copies of the complete book, Rheticus sent several to his colleagues in Nuremberg—certainly Schöner, Osiander, and Petreius—and one to Gasser in Feldkirch. Giese sent a copy to Albert, the duke of Prussia.

The reaction was immediate and emphatic. Gasser wrote to a friend sometime in mid-1540:

> The book may differ from the manner of teaching practiced so far. As a whole it may appear to run contrary to the usual theories of the schools and even to be (as the monks would say) heretical. Nevertheless, what it undoubtedly seems to offer is the restoration—or, rather, the rebirth—of a true system of astronomy. For in particular it makes highly evidential claims concerning questions that have long occasioned much perspiration and debate across the world not only by very learned mathematicians but also by the greatest philosophers: the number of the heavenly spheres, the distance of the stars, the rule of the Sun, the position and courses of the planets, the exact measurement of the year, the specification

of solstitial and equinoctial points, and finally the position and motion of the earth itself.

In August of that year, Johannes Petreius, the Nuremberg publisher, did a most unusual thing. He dedicated a book to Rheticus in the form of a letter, and in that letter he boldly asks Rheticus to convince Copernicus to publish his long-awaited book and to publish it with him, Petreius, in Nuremberg. After congratulating Rheticus on the *Narratio* (calling it a "splendid description"), he states that "I consider it a glorious treasure if some day through your urging his observations will be imparted to us." Petreius hoped that Rheticus would see the dedication as a "kind of reward from us for your labors and study." Then to close the deal, Petreius finishes his letter by reminding Rheticus how Nuremberg is a major trading hub, that his publishing company is able to distribute books to every corner of Europe, and that Schöner, the scholar who had taken such good care of Rheticus during his stay and had taught him much, also wants to see Copernicus published in Nuremberg: "It will fall on you, not only to commend our service, but also to acknowledge and proclaim the great favor of Schöner toward you."

The book that Petreius dedicated to Rheticus was by Antonius de Montulmo, a fourteenth-century physician, entitled *On Natal Horoscopes*. It was on Regiomontanus's Index of Books, and Schöner had discovered the additions to the work that Regiomontanus had planned to include, so it was a very significant book. Dedicating it to the young Wittenberg professor, who was known to be fascinated with horoscopes, represented a serious lobbying effort. Petreius meant business—he was determined to be Copernicus's publisher.

The letters from Osiander, Gasser, and especially Petreius were persuasive. Gasser was a knowledgeable astrologer and published author. He represented the learned community of philosophers whom Copernicus hoped to impress with his work. Petreius's letter was at least as important to the canon because it meant that the astronomers of Nuremberg—which included Schöner and Osiander—all of whom published with Petreius, were also eager to read Copernicus's book. With this level of support, Copernicus finally relented. He would at last publish his entire manuscript, thus allowing anyone who could afford to buy the book or borrow it the chance to study the details of his revolutionary theory of the celestial bodies. On July 1, 1540, he wrote to Andreas Osiander, asking for his advice about how to stifle or at least minimize the opposition to his radical system that was likely to greet his book.

So, starting in the midsummer of 1540, Copernicus and Rheticus went to work preparing the "larger work" for publication.

COPERNICUS'S OLD and dog-eared manuscript was hundreds of handwritten pages long. Over the next twelve months, Copernicus went back through the massive work and made corrections, brought it up to date, and otherwise revised it for publication. He did most of this work himself. One of the three books that Rheticus had given Copernicus the previous year, Regiomontanus's *On Triangles of Every Kind,* was particularly significant, because it caused Copernicus to go back through his trigonometry section and make revisions.

Because the manuscript that Petreius would later work from was written in a hand different from Copernicus's, scholars have long assumed that Rheticus took the finished chapters from Copernicus and redrafted them to create a clean copy for the publisher. (Incredibly, Copernicus's original manuscript has survived the nearly five centuries since its completion. Copernicus never parted from the original; it is now in the archives at the University of Kraków.) While waiting for Copernicus to finish his individual chapters, Rheticus and Zell found a way to occupy themselves—they finished the detailed map of Prussia that had been started many years earlier by Copernicus and Alexander Scultetus. Thus, the two Lutherans were seen traveling all over the region surveying the countryside.

In late 1540, probably November or December, Rheticus returned to Wittenberg to teach the Introductory Astronomy-Astrology course. It was a very short visit, after which it appears that he returned immediately to Frombork.

A couple of distractions intruded on Copernicus's efforts to finish his manuscript. In January 1541, the trial commenced over who had the right to take over Alexander Scultetus's house, vacant since the suspected heretic had fled Frombork several months earlier. Recall that at the trial, which was presided over by a panel of canons, one of the litigants demanded that Copernicus and Niederhoff be forced off the panel because they had been accused of the same offense as Scultetus—that is, Lutheran heresy. The hearing to address the complaint about Copernicus and Niederhoff was put off until April.

Fortunately, Copernicus was able to avoid the embarrassment of the inquiry when in early April he was summoned to Królewiec to attend to the ill George von Kunheim, one of the duke of Prussia's advisors. A new judicial investigation was rescheduled for June. That month came and went without any record of a follow-up hearing.

In July 1541, Achilles Gasser sent Rheticus a copy of the second edition of the *Narratio prima*, which he had personally supervised in Basel. Gasser related in the letter enclosed with the volume that "a stream of requests" was now coming for Copernicus to publish the larger work.

Dantiscus himself received two letters in July from correspondents in Brussels and Louvain urging him to do everything in his power to see to it that Copernicus's book was published expeditiously. The first was from one Cornelius Scepperus, and he stated that the *Narratio* "has made the name of Copernicus famous." The second was from the Dutch scholar Gemma Frisius. In his letter he related to Dantiscus that because of Copernicus and Rheticus, Warmia had become the new intellectual hot spot, "the shelter of the Muses," and their work on the heavens had gained them new admirers.

NICOLAUS COPERNICUS revised the last page of his decades-in-the-making manuscript in late summer of 1541. He had now done everything he could. The rest of the process would be the responsibility of others. Rheticus packed the "fair" copy, the one that he had copied from Copernicus, in his belongings and left Frombork in September, ending a twenty-five-month stay. Many years later he told one of his patrons, "After I had spent about three years in Prussia, the great old man charged me to carry on and finish what he, prevented by age and impending death, was himself unable to complete."

15

The Publication

Rheticus could not miss another semester at the University of Wittenberg. So in late September 1541, he and Zell said good-bye to Copernicus and began their journey back to Wittenberg. Rheticus had spent over two years in Warmia, and before that eight months in the Nuremberg area. Except for his brief return to Wittenberg in the winter of 1540–41, he had been away from his home university for almost three years. Rheticus should have returned as a conquering hero—the author of a much talked-about book, and the lone disciple of a visionary astronomer. And, he possessed the most acclaimed unpublished manuscript in Europe.

His reception in Wittenberg was mixed, though. Back in June 1539, just after Rheticus had arrived in Frombork, Luther said at one of his dinner seminars:

> There is mention of a certain new astrologer who wanted to prove that the earth moves and not the sky, the sun, and the moon. This would be as if somebody were riding on a cart or in a ship and imagined that he was standing still while the earth and the trees were moving . . . So it goes now. Whoever wants to be clever must agree with nothing that others esteem . . . I believe the holy scriptures, for Joshua commanded the sun to stand still and not the earth.

Melanchthon's attitude was probably a little more problematic for Rheticus because he was Rheticus's boss and had been the one to send him away from Wittenberg to become a better astrologer in the first place. Melanchthon wrote to a correspondent in October 1541, just as Rheticus had resumed teaching: "Many hold it for an excellent idea to praise such an absurd matter, like that sarmatic [Polish] astronomer, who moves the earth and lets the sun stand still."

Most of his colleagues, however, must have been happy to have Rheticus back. They recognized the dramatic contribution to astronomy that he had made with the *Narratio prima.* No sooner had Rheticus unpacked than he was told that he had been elected by his fellow professors as their new dean of faculty. The election as dean indicated that Melanchthon still supported Rheticus, even though he disagreed with the science Rheticus had learned at the feet of the "sarmatic astronomer."

Since he was teaching again in Wittenberg, Rheticus was unable to make the long journey to Nuremberg to personally deliver the manuscript for *On the Revolutions* to the publisher Petreius. While in Wittenberg, though, he took the unusual step of taking the two trigonometry chapters from Copernicus's manuscript and publishing them as a separate short book. This book, entitled *On the Sides and Angles of Triangles,* was published in the late winter or early spring of 1542, and it was essentially a complement to Regiomontanus's *On Triangles of Every Kind.* The title page identified the author, in Latin, as "D. [Doctor] Nicolao Copernico Toronensi [from Toruń]." But Rheticus did not just lift the chapters from Copernicus's manuscript. He reworked some of the material and improved it. He dedicated the short book to Georg Hartmann, a Nuremberg mathematician whom he had befriended along with Schöner in

1538–39. In the preface Rheticus announced, "There has been no greater human happiness than my relationship with so excellent a man and scholar as [Copernicus] is. And should my own work ever make any contribution to the general good (to the service of which all our efforts are directed), it shall be owing to him."

When the spring semester ended in May 1542, Rheticus finally left for Nuremberg. He had been holding Copernicus's manuscript for eight months, but once he was in the city of Schöner, the pace picked up immediately. Petreius was given the manuscript and got to work right away. Surprisingly, Rheticus left Nuremberg for Feldkirch to visit his mother and Gasser barely a month later. Perhaps he realized that Petreius did indeed know how to set a technical book.

Meanwhile, the drumbeat of anticipation had already started. Rheticus's colleague at Wittenberg, Erasmus Reinhold, who held the chair in higher mathematics, mentioned *On the Revolutions* in the preface to his new commentary on Peurbach's *New Theory of the Planets*, a much-anticipated book that appeared in 1542. He announced that it was expected that this astronomer from Prussia was the next Ptolemy.

On June 29, a letter was sent by a resident of Nuremberg to a friend, in which he delivered the following news:

> Prussia has given us a new and marvelous astronomer, whose system is already being printed here, a work of approximately a hundred sheets length, in which he asserts and proves that the Earth is moving and the stars are at rest. A month ago I saw two sheets in print; the printing is being supervised by a certain Magister [professor] from Wittenberg.

The same sheets that the writer of the letter saw were also sent to Copernicus, so he knew that work on the book had finally begun. At this point he wrote the last piece to be included in the book, which was his preface and introduction.

THE SUMMER OF 1542 was the last time that everything went well. First, Rheticus accepted a new position at the University of Leipzig. Leipzig was a fine university and his new job represented a promotion to professor of higher mathematics, the job that Reinhold held at the University of Wittenberg. Rheticus also received a substantial raise. In addition, his new boss was Joachim Camerarius, Melanchthon's best friend. So, the move was surely endorsed by the rector. Melanchthon did however caution Camerarius to spell out clearly Rheticus's obligations and salary, leaving nothing to interpretation. Melanchthon had not forgotten the one-year leave that had turned into three years. The new professor of higher mathematics had to depart Nuremberg for Leipzig in mid-October, 1542.

Before Rheticus left, he and Petreius decided that the book still needed an expert overseer. The person asked to assume this role was Andreas Osiander, the theologian and philosopher who had come to know Rheticus in 1538–39, and who had taken such an intense interest in Copernicus and his theory once he read the *Narratio.* He had written or edited five books with Petreius since 1540, so he was a logical choice.

Recall that Osiander had corresponded with Copernicus in 1541, and that Copernicus had specifically asked for his suggestion as to how to minimize the uproar that might await his book. Osiander had answered that one way might be to present the underlying theory in the book as mere hypothesis and essentially to say to the reader, "Do not worry so much about the theory, it's really just the results that matter." Copernicus had rejected this idea. But Osiander still held strong to his opinion. It seems to have been in his character—he was so stubborn in his theological beliefs that he had burned many bridges by this time and was marginalized by the leading Lutherans in Nuremberg. He was certain that a statement like the one he endorsed should begin Copernicus's book.

So Osiander took advantage of his position as the overseer of the publication process and clandestinely slipped in a one-page "To the reader . . ." preface to the book. Unfortunately, it was the first thing that the reader encountered. Also, it was anonymous. Read carefully one could tell that it had not been written by Copernicus, but most readers would have assumed that it was written by the author. Worse, Osiander went too far. It was not incorrect to call Copernicus's theory a hypothesis, but Osiander made other assertions:

> For this art [astronomy] is completely and absolutely ignorant of the causes of the apparent nonuniform motions. And if any causes are devised by the imagination, as indeed very many are, they are not put forward to convince anyone that they are true, but merely to provide a reliable basis for computation . . . Therefore alongside the ancient hypotheses, which are no more probable, let us permit these new hypotheses also to become known . . . So far as hypotheses are concerned, let no one expect anything certain from astronomy, which cannot furnish it, lest he accept as the truth ideas conceived for another purpose, and depart from this study a greater fool than when he entered it.

At about the same time that Osiander was misrepresenting *On the Revolutions*, another disaster struck. Sometime before December 8, 1542, Copernicus suffered a debilitating stroke. On that day, Giese responded to a letter sent by a Frombork canon, "I was shocked by what you wrote about the impaired health of the venerable old man, our Copernicus." His condition was confirmed a few weeks later when Dantiscus responded to Gemma Frisius in the Netherlands, who was very eager to read Copernicus's book, that the canon was now paralyzed. He went on to mention the book itself was being looked after at the publisher's by Rheticus.

THE LONG-LABORED-ON and long-awaited magnum opus of Nicolaus Copernicus finally rolled off Petreius's presses sometime before the end of March 1543. Yet, instead of wild jubilation, the four people closest to the project had horrible surprises awaiting them.

First, Giese. He was so appalled by the anonymous preface by Osiander that he could not enjoy the moment. He later wrote a long letter to Rheticus that reads in part:

> On my return from the royal wedding in Kraków, I found the two copies, which you had sent, of the recently printed treatise of our Copernicus . . . However, at the very threshold I perceived the bad faith and, as you correctly label it, the wickedness of Petreius, which produced in me an indignation more intense than my previous sorrow. For who will not be anguished by so disgraceful an act, commited under the cover of good faith.

Second, Petreius. One of the leading publishers in Europe, he had worked diligently to publish a very complicated book quickly and well. But shortly after the publishing process got started the main supervisor, Rheticus, left for his new position in Leipzig. Then the author himself became physically and mentally incapacitated. Yet, Petreius still got the book out, but among the first responses was the bitter reaction of Giese over the anonymous preface. Giese, a powerful man, wrote a formal letter of complaint to the Nuremberg city council, demanding that they reprimand Petreius and then force him to republish the book. Giese's letter led to a formal hearing in which Petreius was found innocent of any wrongdoing, but which clearly caused him anguish and embarrassment. Petreius wrote an angry reply to Giese that had to be edited by the council to excise the many intemperate passages.

Third, Rheticus. Poor Rheticus suffered two depressing shocks. When he departed from Nuremberg in September or October and left the supervisory process in the good hands of Osiander and Petreius, he assumed that he had nothing more to worry about.

But this account, written many years later, gives some insight into Rheticus's reaction when he excitedly opened his copy of the book:

> Concerning this letter ["To the reader..."], I found the following words written somewhere among Philip Apian's books (which I bought from his widow)... "On account of this letter, Georg Joachim Rheticus, the Leipzig professor and disciple of Copernicus, became embroiled in a very bitter wrangle with the printer. The latter asserted that

> it had been submitted to him with the rest of the treatise. Rheticus, however, suspected that Osiander had prefaced it to the work, and declared that, if he knew this for certain, he would sort the fellow out in such a way that he would mind his own business and never again dare to slander astronomers. Nevertheless, Apian told me that Osiander had openly admitted to him that he had added this as his own idea."

As bad as the first shock was, the second one must have been even more discouraging. After risking arrest and the loss of his job just to visit Copernicus in the first place, then working with the astronomer for more than two years, then writing the potentially controversial *Narratio prima,* then setting up the publication of the big book through his connections in Nuremberg—after all that, when Rheticus opened the finished book, got past Osiander's blasphemous paragraphs, and finally read Copernicus's opening words, his acknowledgments, Rheticus must have been stunned to read that although Copernicus thanked several people, he somehow forgot to thank him. This had to have been a devastating blow to the young mathematician.

Historians of science have been at pains to explain what happened. Several have said that Copernicus was probably just trying to protect the Lutheran professor, who was already in hot water with Luther and Melanchthon. But this explanation is unlikely. First, Rheticus had attached his name to the well-read *Narratio prima*. And he had published Copernicus's short book on trigonometry, in which Rheticus was identified, too. Also, Tiedemann Giese, who knew Copernicus better than anyone else, had no explanation for the oversight. In the letter written in the summer of 1543, after he had finally seen the book, Giese wrote that "your teacher failed to mention you in his Preface to the treatise. I explain this oversight not by his disrespect for you, but by a certain apathy and indifference (he was inattentive to everything which was nonscientific) especially when he began to grow weak. I am not unaware how much he used to value your activity and eagerness in helping him." But Copernicus had been strong enough to remember others in the preface, so this explanation seems unpersuasive.

And fourth, Copernicus himself. What happened? Was Copernicus upset by the separate publication of his trigonometry chapters, containing changes by Rheticus, without his permission? Or, perhaps by the delay of nearly a year before the printing began? Another possible explanation—and this is only speculation—is that after Copernicus observed the acclaim bestowed on the *Narratio prima*, and after the young and enthusiastic Rheticus left Frombork with his masterpiece, Copernicus might have sensed that he would not be around to enjoy the moment of victory, and Rheticus surely would. Perhaps this bothered him so much that he deliberately slighted Rheticus. Whatever the reason, the oversight is glaring.

When *On the Revolutions* rolled off Petreius's press, the canon was lying in his house in Frombork, paralyzed and dying. He may have been sufficiently alert to realize that he would not live to enjoy the acclaim that was likely to greet the first book in nearly 1,400 years to rival Ptolemy's *Almagest*. There must have been many times while he was actively working on the manuscript that he dreamed of the recognition that the book would bring. And he certainly must have felt excitement while preparing it for publication in 1540 and 1541. Now, knowing that the end was near, what despair he must have felt.

* * *

AFTER THE ANONYMOUS and confusing first page written by Osiander, the next item in the book is the letter of 1536 from Cardinal Schönberg, in which he urged the canon to let him publish the work at his (or the Church's) own expense in Rome. Obviously, Copernicus wanted to make it clear that Church leaders were not opposed to what was about to follow. Next, the first words of Copernicus appear, in the "Preface and Dedication to Pope Paul III." In the preface Copernicus boldly asserts his theory that the earth moves around the sun, but he also states that the fear of being "hooted off the stage drove me to almost abandon a work already undertaken." He then thanks Cardinal Schönberg and Tiedemann Giese, but not Rheticus.

The astronomer proceeds to state that the main problem with past theories and the reason why he tried to come up with a new model was that the other ones "contradict the first principles of regularity of movement." Then he provides a great metaphor, perhaps drawing on his training as a doctor: he compares past theories as consisting of all the parts of a body, "hands, feet, head, and the other limbs"—but put together in such a way that the result is a "monster rather than a man." The author's new conception makes the parts fit as a whole. He goes on to point out that only mathematicians will really be able to pass judgment on what follows.

On the Revolutions consists of six "books." The books are composed of many short chapters. The work is carefully organized, and Copernicus took pains to provide good transitions, introductions, conclusions, and passages meant to help the reader know what has already been covered and what is coming next. But the book is unapologetically technical, with page after page of math, numerous complicated drawings, and many dense tables of numbers.

Book One is a general introduction to the model. Copernicus first discusses the importance of astronomy and then he begins his presentation. The universe is a sphere, as is the earth, the movements of the celestial bodies are regular and circular, and the earth, too, has a circular motion. All of the heavenly bodies move with uniform speed, which is a critical component of his model. Very early in the book, no more than twenty pages in, he describes the "movement of the earth"—that is, that our planet rotates once every twenty-four hours, that it revolves around the sun once every 365 days.

Book Two discusses the rotation of the earth itself and the angle of inclination of the axis. Within this section, the author describes how to construct an astrolabe, which is used to examine the position of the moon and stars. Copernicus points out that the earth's rotation and revolution are slow and natural, and that is why the planet does not break apart and the atmosphere does not blow away. Book Three addresses the movement of the earth around the sun. The remaining three books describe the movements of the moon and the other planets.

THERE WAS NO DOUBT that Nicolaus Copernicus's book was a remarkable achievement. That much was obvious to almost every reader. But, because of its complexity, not much else was. Most interested readers would need some help to understand its implications.

16

THE DEATH OF COPERNICUS

ON DECEMBER 8, 1542, Tiedemann Giese wrote back to canon George Donner, who had informed him about Copernicus's stroke, to express his shock. He continued:

> Just as he [Copernicus] loved his privacy while his constitution was sound, so, I think, now that he is sick, there are few friends who are affected by his condition. Yet we are all indebted to him for his uprightness and outstanding learning. I know, however, that he always reckoned you among those most faithful to him. I therefore ask you, if his condition so warrants, please to watch over him and take care of the man whom you cherished at all times together with me. Let him not be deprived of brotherly help in this emergency. Let us not be considered ungrateful toward this deserving man.

Giese was beseeching a new canon to watch over Copernicus. Where was everyone else? Rheticus had been gone from Frombork for over a year. Giese lived about seventy miles away. Anna Schilling now lived in Gdansk, fifty miles away. Scultetus was in Rome and in deep trouble. Reich was already dead. Copernicus was probably in his large house outside the walls of the cathedral, and at least he had his two servants to assist him.

Of course, his last months were also the difficult winter months of the Baltic region.

On the eve of the publication of his magnum opus, Copernicus was at the lowest point of his life. Knowing that he was dying and would probably never leave his bed again, he was completely isolated. At least he had been well enough during the period after the publication of the *Narratio prima* to know that learned Europe was confident that the canon of Frombork had produced something monumental.

On May 24, 1543, at the age of seventy, Nicolaus Copernicus took his last breath. Giese reported his death to Rheticus: "This was caused by a hemorrhage and subsequent paralysis of the right side on 24 May, his memory and mental alertness having been lost many days before. He saw his treatise only at his last breath on his dying day." If this account is accurate, and there is no reason to suppose that it is not, the published copy of *On the Revolutions* arrived at his house in Frombork on his last day of life. It appears that Copernicus had the will to live to see the book that contained his lifetime commitment to understanding the heavens, the work that in the years ahead would make his name one of the most revered in Western history.

THE FATHER OF ASTRONOMY was buried inside Frombork Cathedral, near the main altar. His remains rested near those of his uncle Lucas. On June 1, just a week after his death, Copernicus's tower was assessed at being worth 30 marks, and the *curia* just outside the walls at 100 marks. Canon Achacy von Trenck purchased the tower right away. It appears that the turret was later called "turris Copernici." One month later, Leonard Niederhoff purchased the *curia*. Either the economy deteriorated or Niederhoff did not maintain the *curia,* because when he died in late 1545, the value had declined to 90 marks. Neiderhoff also purchased Copernicus's countryside grange (*allodium*), which was called "Sebleck number two." After Niederhoff's death, Johann Zimmerman bought the grange.

Copernicus had left a last will and testament, and the executors were Dietrich von Rehden, Leonard Niederhoff, George Donner, and Michael Loitz (who took over Copernicus's canonship). He left his manuscript for *On the Revolutions* to Giese, who in turn bequethed it to Rheticus. Copernicus left his wealth to his nieces Christine and Regina, who were his sister Catherine's daughters, and the seven children of Regina. All his years as a canon, collecting income primarily from the peasants of Warmia, left Doctor Nicolaus with a substantial estate worth more than 500 marks.

On September 10, the chapter wrote to Dantiscus about Anna Schilling. They reported that "from time to time [she] comes here and stays several days for the sake of taking care of her property . . . We are not sure whether she will be able lawfully to be prohibited from coming here, since the legal obstacle is inoperative [meaning Copernicus was no longer alive]. For when the cause is removed, so is the effect."

In this letter, the canons mention that the matter was tried in Dantiscus's court, which means that Anna's banishment had been made official in a court of law.

Immediately upon receiving the letter, Dantiscus (in a response cited earlier) wrote back to say that Anna was not allowed back in Frombork:

> For it must be feared that by the methods by which she deranged him [Copernicus], who departed from the living a short while ago, she may take hold

> of another of you, my brothers . . . I would consider it better to keep at a rather great distance than to let in the contagion of such a disease. How much she has harmed our church is not unknown to you, my brothers, for whom I hope happiness and health.

Except for Rheticus, most of those who played a direct or indirect role in the drama of Copernicus's last years followed him to their graves shortly afterward. Martin Luther passed away in 1546. Johann Schöner, in 1547. Bishop Dantiscus died at the age of sixty-three in 1548. Giese, who succeeded Dantiscus as Bishop of Warmia, died just two years later, in 1550. Both churchmen, Dantiscus and Giese, were buried in Frombork Cathedral. And the publisher Petreius saw his last book come off his presses in 1550. Melanchthon would live until 1560. It was up to Rheticus to direct the Copernican Revolution.

17

Rheticus after Copernicus

In the fall of 1542, Rheticus was about to begin his new position at the University of Leipzig, one that would pay him significantly more than the other professors there, and much more than he had been paid by the University of Wittenberg. He was a celebrity among astronomers thanks to his writing of the *Narratio prima,* and his well-known role in the preparation of *On the Revolutions*. Only twenty-eight years old, he had impressed the finest publisher in Nuremberg, Petreius; the top astrologer, Schöner; and one of the leading theologians, Osiander. There must have been days when he pinched himself over all that had transpired since his first visit to Nuremberg in the fall of 1538, just four years earlier. And there was every reason to suppose that this was just the beginning of a brilliant career. The publication of *On the Revolutions* would send shock waves throughout Europe, and with Copernicus in failing health, Rheticus would be the leader of the Copernican Revolution. Rheticus could well be the next Regiomontanus.

It didn't happen.

Although he lived for another thirty-two years, and had several moments of acclaim, he never came close to equaling the astounding four years between 1538 and 1542, especially the two years between 1539 and 1541.

* * *

In the same letter (dated July 26, 1543) in which he apologized for Copernicus's unforgivable oversight, Giese tried to come up with a plan to force Petreius to issue a new printing of *On the Revolutions.* In place of the note to the reader by Osiander, Giese strongly urged Rheticus to include the short biography of Copernicus that Rheticus had written (and that Giese had already read) along with another piece that Rheticus had prepared that explained why the heliocentric system did not contradict Holy Scripture. Giese said: "In this way you will fill the volume out to a proper size and you will also repair the injury that your teacher failed to mention you in his Preface to the treatise." Giese's entreaty revealed that Rheticus had more material about Copernicus and his system to share with the scholars of Europe. Rheticus did not heed Giese's instructions. If he had, there might have been a way to find another publisher, perhaps in Gdansk, who would have published a new edition of Copernicus's work. Giese could probably have secured a subsidy for that publisher, too, if the book was considered too expensive to produce.

Rheticus's longtime mentor and friend Achilles Gasser also issued a plea for a new volume. In the preface to a book published in 1545, Gasser made it clear that the astronomy-astrology community needed a second book, one that would be easier to understand than the original:

> [S]ince the demise of that most illustrious gentleman Nicolaus Copernicus, your teacher, it is now up to you alone . . . So for us uncultivated people having scant aptitude and no Theseus to lead us up the steep terrain, offer an easier introduction to this science, as well as clearer proofs. But above all may

> you strive to bring forth tables accessible to those of common capacity . . . It remains then for you to set forth with great clarity how this business is to be most carefully and skillfully written up.

Gasser's public appeal in the form of a preface had surely been made in private between the two men beforehand, probably soon after Gasser bought his copy of *On the Revolutions* and realized how difficult it was.

If Gasser felt this way, many others must have as well. However, Rheticus did nothing. Why did he not respond to pleas by two of the most important people in his life, men whom he had listened to in the past? Had he been too offended and embarrassed by Copernicus's lack of gratitude? A few years later, Rheticus would again describe Copernicus in glowing terms. But for now he sat on his hands.

RHETICUS LEFT FOR LEIPZIG in September 1542. He was now professor of higher mathematics and astronomy (he had been professor of lower mathematics at the University of Wittenberg). He taught at the University of Leipzig for three full academic years. During this incredibly important period for the nascent Copernican Revolution, he did nothing except teach classes.

In the fall of 1545, he took another leave from teaching and traveled to Milan to seek out another celebrity scholar—the astrologer Girolamo Cardano (1501–1576). Rheticus remained as fascinated with astrology as ever, and Cardano was easily the most famous astrologer in Europe. His fame rested largely on several best-selling books that he had published with Petreius. It appears that Rheticus was not overly impressed with Cardano or his Italian colleagues, and he left Milan disappointed.

After being absent for a full year, Rheticus received a letter from his dean at Leipzig informing him that he needed to return to his post. He responded by asking for a raise and then proceeded to stay away from Leipzig for another two years. During this period he traveled from town to town in the region of his childhood, and it appears that for an extended period he suffered from something akin to a nervous breakdown. At one point he was bedridden for five months in the Swiss town of Lindau, not far from his hometown of Feldkirch and near his mother and stepfather. References to him in letters indicate that he was deeply troubled. It also seems that he had a form of religious conversion, and became what today we would call "born again."

Rheticus finally returned to Leipzig in the late summer of 1548, after a three-year absence. He made a successful recovery from whatever had been afflicting him because the next brief period of his career was quite productive. He plunged into teaching again, but, more important, he found his writing voice once more. Over the next three years he produced a new edition of Euclid's *Geometry*, an important book on trigonometry, and an almanac. The almanac was significant because he announced that it was based on the astronomy of Copernicus, the first time since his work on *On the Revolutions* that he publicly embraced Copernicus.

THE INDIFFERENCE of Copernicus had been hurtful and deflating. His interaction with Cardano had been disappointing. His breakdown in the region of his hometown had been unsettling. However, it looked like Rheticus had finally put these trials behind him and recovered. Then disaster struck. In April 1551, Hans Meusel, a merchant, brought a lawsuit against Rheticus for a shocking crime—the injunction claimed that the professor had "lure[d] my son . . . plied him with strong drink, until he was inebriated; and finally did with violence overcome him and practice upon him the shameful and cruel vice of sodomy."

Rape, then as now, was a profoundly serious offense. Joachim Rheticus fled Leipzig immediately, leaving nearly all of his personal belongings behind. Soon he appeared in Prague, ostensibly to study medicine.

Over the next twelve months official letters were sent from the court to Rheticus and ignored. But on April 11, 1552, Rheticus, age thirty-eight, was found guilty of raping young Meusel. He was exiled from Leipzig for 101 years. Among the many things he left behind in Leipzig was a mountain of debt, equal to twice his annual salary.

IN 1554, Rheticus emerged again, this time in Kraków, practicing medicine. At about the same time, he was apparently offered a job at the University of Vienna, the institution of Peurbach and Regiomontanus, where exactly a hundred years earlier Peurbach had taught his famous astronomy course. Rheticus did not take the position.

In Kraków, he built a 50-foot-high obelisk, which was used extensively in his trigonometric work. It appears that Rheticus formed few friendships, and certainly no intellectual partnerships, in the Polish capital.

In 1566, the second edition of *On the Revolutions* was published in Basel. It included Rheticus's *Narratio prima.* His name was prominently displayed on the title page, thus raising his profile.

Still restless, Rheticus left Kraków in 1572 for a small town in Hungary called Cassovia. At the same time a student at the University of Wittenberg, Valentin Otto, had just discovered geometry and trigonometry, and read Rheticus's 1551 publication, *The Canon of the Science of Triangles.* Eerily, just as Rheticus had felt thirty-five years earlier that he simply had to find and confer with Nicolaus Copernicus, Otto decided to seek out Joachim Rheticus. So in the spring of 1574, Otto journeyed to Cassovia. Otto later described the moment when he first met Joachim Rheticus:

> When I returned to the University of Wittenberg, fortune willed that I should read a dialogue by Rheticus, who had been attached to the Canon. I was so excited and enflamed by this that I could not wait but had to journey at the first opportunity to the author himself and learn from him personally about these matters. I went, therefore, to Hungary where Rheticus was then working and was received by him in the kindest manner. We had hardly exchanged a few words on this and that when, on learning the cause of my visit, he burst forth with the words: "You come to see me at the same age as I myself went to see Copernicus. If I had not visited him, none of his works would have seen the light.

Rheticus was actually sixty; Copernicus had been sixty-six when Rheticus first met him. Rheticus died of a respiratory infection, possibly pneumonia, about six months later on December 4, 1574.

Fortunately, much as Rheticus performed heroically on behalf of Copernicus, Otto secured Rheticus's legacy in the field of trigonometry. Twenty-two years later, in 1596, Rheticus's *Opus palatinum de triangulis,* which provided the most complete trigonometric tables to date, was finally published, thanks to Valentin Otto.

18

The Impact of *On the Revolutions*

The first printing of *On the Revolutions* was only about four hundred copies, which was a standard print run for a technical book in the sixteenth century. It was very expensive—about 1 florin. To put that price in perspective, Rheticus made 100 florins a year as a professor at the University of Wittenberg. If it is difficult to imagine that only four hundred copies of a book changed the world, recall that an equally important but more accessible book, Charles Darwin's *On the Origin of Species*, sold only 4,250 copies in its first few years. Copernicus's work was intended for the relatively small number of well-trained mathematicians and astronomers in Europe. Copernicus and his publisher, Petreius, never expected the book to be a best seller.

The scholar most responsible for the immediate positive impact of *On the Revolutions* was not Rheticus, but rather his former colleague and perhaps rival at the University of Wittenberg, Erasmus Reinhold (1511–1553). Reinhold was three years older than Rheticus, and they had been students at the University of Wittenberg at the same time. They were both hired as professors in the summer of 1536 by Melanchthon, but

Reinhold received the more desirable of the two appointments, professor of higher mathematics. In 1542, he published a new commentary on Georg Peurbach's *New Theory of the Planets*; it was a highly coveted work within the astronomy-astrology community. In the preface of the book, Reinhold announced his excitement about the impending publication of Copernicus's great book, which he probably learned about from Rheticus or his friends: "I know of a recent scholar who is exceptionally skillful. He has raised a lively expectancy in everybody. One hopes that he will restore astronomy . . . I hope that this astronomer, whose genius all posterity will rightly admire, will at long last come to us from Prussia . . ."

In 1549, Reinhold published a new commentary on Ptolemy's *Almagest*, and this title also had a significant impact in astronomical circles. Then, in 1550, he succeeded Melanchthon as rector of the University of Wittenberg.

Reinhold immersed himself his copy of *On the Revolutions* and was the first to prepare something practical from it. Copernicus's work was largely theoretical, and many readers—mainly the astrologers—simply wanted to utilize the data denoting the planetary positions. So, Reinhold gave them precisely what they coveted—astronomical tables based on Copernicus's calculations. Published in 1551, they were called the Prutenic Tables (Prutenic after Prussia). The volume became a sixteenth-century best seller and enhanced the reputation of Copernicus significantly. Reinhold wrote: "Therefore we are really indebted to the great man, Nicolaus Copernicus, for having generously communicated to the scholars the observation of several years of watches, and for having, by his assiduous work, given a new birth to the doctrine of the movements, which is enlightened by his publication about the Revolutions." Strangely, Reinhold did not care about Copernicus's overall conception of the universe on which the numbers were based, and ignored the larger issue—heliocentrism versus geocentricism. Reinhold *was* very impressed, though, that Copernicus was working toward a universal law—uniform circular motion of the planets—and that he had striven to eliminate Ptolemy's equant from his model of the heavens.

Like Regiomontanus before him, Reinhold died of the plague. He was only forty-two, and he passed away just two years after his tables were published, in 1553. This was another huge loss for the astronomy community. Reinhold had prepared a positive commentary on *On the Revolutions*, but his death prevented it from being published.

For several decades, the scholarly community's reaction to Copernicus's book was similar to Reinhold's—it wanted the planetary tables and data, but did not pay a great deal of attention to the "earth in motion" model. Copernicus became well known primarily as an astronomer who did heroic work and left behind more accurate calculations of the night sky. It is probable that Osiander's anonymous "To the reader . . ." had a tangible impact on how readers viewed the book. For instance, Johannes Stadius (1527–1579), a student of Gemma Frisius's in the Low Countries, used *On the Revolutions* to prepare yet another set of astronomical tables. Published in 1556, they were easier to use than Reinhold's, but again there was no mention of the arrangement of the heavens from which the tabulations were derived. (One reader of Stadius's new tables was Nostradamus, who died in 1566 and wrote the words "death is close at hand" in his copy of Stadius's book.)

JUST AS REGIOMONTANUS LEFT behind no equally talented heir, the same situation arose after the death of Copernicus.

Though Rheticus could have been his successor, he did not rise to the occasion. Neither did Reinhold, who died prematurely. The first of the great successors arrived two generations later, in the person of Tycho Brahe (1546–1601), who was born three years after the Canon of Frombork passed away.

Brahe was born into Danish nobility, the oldest son of eleven children. He spent his teens and early adult years attending several universities, similar to the way Copernicus spent the last years of the fifteenth century. He also had a connection to the University of Wittenberg—he attended classes there for five months in 1566, and studied astronomy under Reinhold's successor and Melanchthon's son-in-law (and pro-Copernican), Caspar Peucer. Later in 1566, while studying at the University of Rostock, Brahe fell into an argument with another student that escalated into a duel. In the ensuing sword fight, Brahe's foe sliced off his nose, and the Dane had to wear a gold nosepiece for the rest of his life.

Early in his schooling, Tycho fell in love with astronomy. He started buying astronomical instruments and books (including the *Alfonsine Tables,* which was one of Copernicus's first purchases at the University of Kraków), and making observations. Surveying the night sky at every opportunity, Brahe knew it like the back of his hand. Back in Denmark, on November 11, 1572, Tycho made one of the most significant observations in the history of astronomy. A little after sunset on that day, he noticed an extraordinarily bright star that he knew had not been there the night before. Brahe went on to observe it for the next sixteen months, until it was no longer visible in the firmament. It is now known that he observed a supernova, or exploding (dying) star. This bright light was seen by all astronomers in Europe, most of whom thought that it was a comet, then believed to be phenomena that occured between the moon and the earth. However, only Brahe, who had the stamina of a bull, took careful enough measurements to know that it was a star, a fixed star.

Thus Brahe alone recognized the significance of the bright new light in the sky. How could there be a new star in the immutable and perfect heavens made by the Creator? Clearly, the heavens were not unchanging. This insight cast further doubt on the prevailing Ptolemaic orthodoxy. In 1573, Tycho published a short book, only about fifty pages long, about his discovery. The book and its author quickly became famous.

The ambitious Tycho then used his newfound status to persuade the king of Denmark to build him an observatory. The king gave Brahe a small island near Copenhagen, called Hven, and paid for what quickly became the finest astronomical center in Europe, one filled with the most advanced instruments available. The king also provided enough financial support for Tycho to hire dozens of research assistants. Brahe, already rich through his family, became even wealthier under the king's patronage. Like nearly all astronomers before him—except for Copernicus—Tycho believed strongly in astrology. He was an active astrologer, producing prognostications and horoscopes for his patrons, including the king of Denmark, for most of his career.

Brahe was fascinated with Copernicus. Though he was never won over to the heliocentric model, he nevertheless honored its founder's memory. He worked diligently to obtain an already rare copy of the *Commentariolus* (which Brahe is credited for naming, by the way; until Brahe, it had no formal title). As mentioned earlier, he actually sent an assistant to Frombork in 1584 in order to gain a better understanding of Copernicus's observations. Elias Olsen Morsing, the assistant, stayed with the canons on Cathedral Hill making observations for an entire month. They gave him Copernicus's triquetrum, which the canons had guarded with care for over forty years, to give to Tycho. Tycho cherished it for the rest of this life, and displayed it in what was essentially a museum on his island.

On November 13, 1577, Brahe added to his fame. That night he saw for the first time a large comet with a long tail. After observing it carefully for about two months, the Dane knew that it was not a sublunar (occurring between the earth and the moon) phenomenon. It was out in space, among the planets, which meant that it must be piercing the planetary spheres. Recall that those who adhered to Ptolemy's model believed that the planets were attached to physical, tangible spheres—that is, spherical shells. Here was proof that there were no shells. The 1572 supernova showed that the heavens were changeable, and the comet proved that the planets did not ride on celestial spheres. What else about the old model was not valid? Tycho wrote about his findings in another short book, this one in German and intended for all scholars, not just mathematicians. The already famous Dane gained even more acclaim and was now the undisputed king of astronomy of his generation.

Shortly after the observations of the comet of 1577–78, Tycho developed his own model of the heavens. He called it his Tyconic system, and it was essentially a hybrid of the systems of Copernicus and Ptolemy. His geoheliocentric system called for a perfectly still Earth (it did not rotate) to be the center of the universe, with the planets (Mercury, Venus, Mars, Jupiter, and Saturn) revolving around the sun. The sun, moon, and fixed stars revolved around the earth, as in Ptolemy's conception. Brahe never fleshed out his model the way Copernicus did, but he described it in a book published in 1588, *De mundi aetherei recentioribus phaenomenis.*

Tycho said: "What need is there, without any justification, to imagine the earth, a dark dense and inert mass, to be a heavenly body undergoing even more numerous revolutions than the others, that is to say, subject to triple motion, in violation not only of all physical truth but also of the authority of Holy Scripture, which ought to be paramount."

Frederick II, Tycho's patron, died in 1588. The new king, Christian IV, was less committed to the astronomer's research program, which was quite expensive to maintain. Over the next decade Brahe slowly fell out of favor with the young king and had to close down his observatory. He left his island in 1597. Brahe eventually settled in Prague, but his glory days were behind him. He made one more major discovery, though—Johannes Kepler, who became his research assistant in early 1600.

Johannes Kepler (1571–1630) was a prodigy. His teachers recognized his impressive mathematical abilities early in his schooling. At the age of twenty, he received his master's degree from the University of Tübingen, where he studied under Michael Maestlin (1550–1631), who was an early (though stealth) Copernican. Like Reinhold, Maestlin had carefully studied *On the Revolutions* and thought deeply about its implications. The young professor discussed Copernicus among his students but did not formally teach heliocentrism.

Kepler was a deeply committed Lutheran who had hoped to become a minister and theologian, but he was compelled instead to teach mathematics and astronomy at a secondary school in the town of Graz (which is now in Austria). Kepler gained local fame in 1595, when the prognostications in his astrological calendar for that year turned out to be remarkably accurate, especially his prediction of an invasion of Europe by the Turks.

Though Tycho deeply respected Copernicus and *On the Revolutions*, Kepler was the first true Copernican after Rheticus. As he later related:

> When I was studying under the distinguished Michael Maestlin at Tübingen six years ago, seeing the many inconveniences of the commonly accepted theory of the universe, I became so delighted with Copernicus, whom Maestlin often mentioned in his lectures, that I often defended his opinions in the students' debates about physics. . . . I have by degrees—partly out of hearing Maestlin, partly by myself—collected all the advantages that Copernicus has over Ptolemy.

This passage appeared in his important book, the first of many for the prolific Kepler, entitled the *The Cosmic Mystery* (*Mysterium cosmographicum*). In its pages, Kepler presented his model of the universe, which was decidedly Copernican. The goal of the effort was to show with first principles drawn from geometry why the universe was configured as it was, with the sun at the center, the earth and the other planets in motion around it, and the planets in their specific locations and distances from the sun. He also argued forcefully for the sun being the reason for this configuration. Copernicus placed the sun at the center of the universe but did not give it an active role in the scheme—Kepler did, thus anticipating Newton and the theory of universal gravity.

Like Copernicus and Rheticus, Kepler had no problem reconciling scripture and the heliocentric theory. For all three,

God's greatness and benevolence were evident in the extraordinary beauty and harmony of this system. Kepler expressed his attitude to Maestlin, who would remain a lifelong friend and correspondent: "I wanted to become a theologian . . . for a long time I was restless. Now, however, behold that through my effort God is being celebrated in astronomy."

Kepler's *Cosmic Mystery* was published in Tübingen, where the professors at the university and his publisher had asked him to make part of his book understandable to laymen. So Kepler wrote a very accessible introduction to his work that directly addressed why Copernicus's system was dramatically better than Ptolemy's. The book was published in 1596, fifty-three years after *On the Revolutions*. From then on, the tide turned toward Copernicus's model of the heavens. Kepler sent copies of the book to Galileo (whom he had never met) and Tycho (whom he had also never met). Although the book contained several fundamental flaws, they were not discovered until later. The work had a powerful influence. A leading scholar of early astronomy once noted: "Although the principal idea of the *Cosmic Mystery* was erroneous, Kepler established himself as the first . . . scientist to demand a physical explanation for celestial phenomena. Seldom in history has so wrong a book been so seminal in directing the future course of science."

With the Counter-Reformation gathering steam, the staunchly Lutheran Kepler and his family felt the need to leave Catholic Graz in 1599. He made a bold move to go directly to the most famous astronomer in Europe—Tycho Brahe, who had recently relocated to Prague. Kepler was already quite distinguished in astronomical circles, but Brahe treated him as he did all of his assistants, subserviently. Like everyone else, Kepler was given a specific task. He was assigned to focus only on the movements of the planet Mars. Kepler quickly became dissatisfied with his relationship with the difficult Brahe and was getting close to moving on to a new post, when Brahe fell terminally ill. Only fifty-six years old and with mountains of work to finish, the Dane beseeched Kepler to continue the important work on his new astronomical tables. It was a scene that harkened back to Peurbach urging Regiomontanus to continue with the *Epitome*, Copernicus trusting Rheticus with his life's work when he left Frombork, and Rheticus charging Otto to complete his trigonometry volume. Brahe called his tables the Rudolphine Tables after Emperor Rudolf II, his patron in Prague. The deeply religious Kepler thought that God himself had intervened: "God let me be bound with Tycho through an unalterable fate and did not let me be separated from him by the most oppressive hardships." Brahe expected Kepler to make the tables fit with his geoheliocentric theory, but it did not work out that way.

Kepler was troubled by certain discrepancies in the orbits of the planets. He knew that they could not be ascribed to measurement errors because he was aware of just how accurate Brahe's measurements were. Though there was only an eight minutes of arc inconsistency with perfect circular orbits, those eight minutes could not be ignored. There had to be some way to explain it. By 1605, after great effort, Kepler devised the idea that carved his crucial place in the history of science—the ellipse: "With reasoning I derived from the physical principles agreeing with experience, there is no figure left for the orbit of the planet except a perfect ellipse." Copernicus, and all astronomers, had believed that the only possible shape for an orbit was a perfect circle or sphere, since God the creator would build only with a perfect shape. For the deeply religious Kepler to abandon this belief was remarkable. His first book based on his profound insight was *Astronomia nova*, published in 1609.

Moving often for professional and religious reasons, and authoring many other works (all the while charged with finishing Tycho's tables), Kepler found time to write the definitive technical book that took Copernicus's fundamental insights in *On the Revolutions* and redid them using his own discovery of elliptical orbits. The book was entitled *Epitome astronomiae Copernicanae* and it appeared in three volumes from 1617 to 1621. Kepler's *Epitome* was a treasure, and it was later utilized by Isaac Newton.

Active and constantly writing to the end, Kepler died of a fever at the age of fifty-nine in 1630. He had finally finished Tycho's tables and had them published in 1627.

JOHANNES KEPLER firmly established the fundamental truth of Copernicus's heliocentric model among astronomers and mathematicians. But Kepler's published works, like Copernicus's, were unrelentingly technical. The scholar who established the Copernican view among laypeople was his contemporary Galileo.

Galileo Galilei (1564–1642) was born seven years before Kepler. The native of Pisa was the oldest of seven children born to a well-to-do family; his father was a noted musician and music theorist. Galileo was first attracted to mathematics, and he became a professor of mathematics at the University of Pisa in 1589, at the age of twenty-five. In addition to mathematics, he was also fascinated by basic physics. He appeared to be indifferent about astronomy, but when Kepler sent him a copy of his *Cosmic Mystery*, Galileo's letter of thanks revealed a deep interest:

> Many years ago I accepted Copernicus's
> theory, and from that point of view I discovered

> the reasons for numerous natural phenomena, which unquestionably cannot be explained by the conventional cosmology. I have written down many arguments as well as refutations of objections. These however, I have not dared to publish up to now. For I am thoroughly frightened by what happened to our master, Copernicus. Although he won immortal fame among some persons, nevertheless among countless (for so large is the number of fools) he became a target of ridicule and derision. I would of course have the courage to make my thoughts public, if there were more people like you. But since there aren't, I shall avoid this kind of activity.

By Galileo's time there were many who thought that Copernicus had been subject to "ridicule and derision" during his lifetime. It was only after his death that this occurred. Although Galileo says in this letter that he wants to avoid public ridicule he later showed great bravery in the name of the truth that he believed in.

Galileo was famous in Padua, where he became a professor of mathematics in 1592. When in 1604 another supernova was visible, Galileo gave public lectures about the new star before large crowds at the university. A confirmed anti-Aristotelian, Galileo argued that the new star showed that the heavens were not perfect and unchangeable, precisely the same argument that Tycho had made thirty years earlier in his short book.

Galileo built a telescope soon after the first telescope was made in Holland. Galileo's was more powerful, and in January 1610, he started doing research with his new instrument. He immediately discovered that the moon had craters and mountains, and that there were many more stars than previously thought. But most significantly, he saw four moons orbiting Jupiter. Galileo immediately wrote a short (sixty page), nontechnical manuscript describing these discoveries. *The Starry Messenger* (*Sidereus nuncius*) was published almost instantly. It was available by March. The book sold quickly, making Galileo a famous man throughout Europe almost overnight. Within months, Kepler publicly endorsed Galileo's results. The moons around Jupiter confirmed once and for all that not all heavenly bodies revolved around the earth, essentially killing forever the Ptolemaic cosmology.

Galileo left Padua to return to the University of Pisa. In 1613, he wrote a second short book about another telescope-based discovery. Entitled *Letters on Sunspots*, this was Galileo's first formal endorsement of the Copernican system. Almost immediately the book got him into trouble with the church hierarchy. In 1616, Galileo went to Rome to attempt to make his views on science clear. Though not as devout as Kepler, Galileo was nonetheless religious. He believed that scripture and the Church's teachings addressed moral issues, and that science and nature were separate. Galileo's most famous book came out in 1632, *The Dialogue Concerning the Two Chief World Systems*, which pitted an Aristotle-Ptolemy supporter against a Copernican as they tried to win over a young uncommitted student to their respective points of view. It was not militantly Copernican, though in the end the Copernican system won the day. The narrative was very readable, and it became another highly successful book for Galileo. It was probably too successful, in fact.

Galileo was ordered to appear before the Inquisition in October 1632. Pope Urban VIII found several passages in the *Dialogue* to be direct affronts to positions that he held; the pope's beliefs were endorsed in the book by the buffoonish character called Simplicio. The trial began in February 1633, and Galileo was quickly found guilty of voicing the Copernican worldview, which had been officially outlawed since 1616, when *On the Revolutions* was placed on the Index of Forbidden Books by the Congregation of the Index. Galileo was sentenced to house arrest for the rest of his days. Though the sentence was not onerous, and he was able to do practically anything he wanted, the effect was severe. Galileo was never the same. His eldest daughter, to whom he was very close, died suddenly in 1634. Galileo followed her to the grave in 1642.

On the Revolutions was easily available throughout its time on the Index, but the ban on openly endorsing the Copernican cosmology was sufficient to accuse Galileo of heresy. Copernicus's book was not removed from the Index for more than two hundred years—the first Index on which *On the Revolutions* did not appear was published in 1835.

Kepler and Galileo established the heliocentric model, but Isaac Newton added an important element when he proved the existence of universal gravity. Finally it was understood why the planets orbited the larger sun, and why the small moons around Jupiter and the earth orbited their home planets.

The daily rotation of the earth on its axis was proved only in 1851, when Jean-Bernard-Léon Foucault (1819–1868) set up his famous pendulum. When the pendulum circled clockwise, returning to its original spot after twenty-four hours, the rotation was confirmed.

* * *

As Copernicus was working with Rheticus in 1540–41, he must have known that he was polishing a special and important book. We know that he saw it as a volume primarily for astronomers, one that would help with reforming calendars and producing yearbooks that would provide farmers with improved information about the planting seasons. And, he knew, it would assist astrologers, too. He must have taken a quiet pride in knowing that the sheer breadth and rigor of the book were equaled by only one other, Claudius Ptolemy's *Almagest*. But, it is safe to say that he could have had no idea that it would be a powerful catalyst for the science that over time would usher in the modern world.

Copernicus did two things that cannot be overestimated. First, he bravely (not because of fear of imprisonment, but because of fear of ridicule) presented a theory that aggressively contradicted the greatest Greek philosopher, Aristotle; the greatest astronomer, Ptolemy; and centuries of popular belief endorsed by the Church, and he did not flinch. He stayed true to his unique vision. Second, he sought to work from first principles. His principles were perfect uniform motion and perfect circles. This effort to work from first principles appealed to the scholars who embraced his work and extended it in the centuries to come, including Brahe, Kepler, and Galileo.

Nicolaus Copernicus was the starting point; he was the mold breaker. When Newton discovered universal gravity, everything made sense, and there was no stopping modern science. But it all started with one obscure man who lived in a backwater place, worked by himself with primitive instruments, and labored as few had before him for no direct gain. He was an original, a unique mathematical talent. One hopes that he had at least a glimmer of the ultimate impact of his achievement as he lay on his bed after his stroke in the winter of 1542–43, waiting to meet his God, whose beautiful creation he had seen more clearly than any human before him.

Notes and Select Sources

NOTES

Chapter One Prelude to Future Troubles

1 *"My lord, Most Reverend Father"* Letter from Copernicus to Bishop Maurice Ferber, July 27, 1531, in Rosen, *Copernicus and the Scientific Revolution,* pp. 149–50; this letter also appears in Rosen, *Minor Works*, pp. 319–20.

3 *"Copernicus was a late bloomer"* The rest of this chapter reflects the consensus on the basic facts of Copernicus's life. The standard sources are in fundamental agreement; those English sources are Rosen, Swerdlow, Biskup, Gingerich, and Westman (full citations are given in Select Sources).

8 *a manifesto for the heliocentric theory* What follows is based on the *Commentariolus* by Nicolaus Copernicus, in Rosen, *Three Copernican Treatises,* pp. 57–90.

10 *"I had kept [the manuscript] hidden"* Copernicus, *On the Revolutions*, pp. 8–9.

Chapter Two The Precursors

12 *if not for their untimely deaths* This chapter is based on the following sources for the lives and works of Peurbach and Regiomontanus: Donald deB. Beaver,

"Bernard Walther: Innovator in Astronomical Observation," *Journal of the History of Astronomy;* C. Doris Hellman and Noel Swerdlow, "Peurbach, Georg," *Dictionary of Scientific Biography;* Barnabus Hughes, translator, Regiomontanus's *On Triangles of Every Kind;* Olaf Pederson, "The Decline and Fall of the *Theorica Planetarum,*" *Studia Copernicana;* Edward Rosen, "Regiomontanus, Johannes" *Dictionary of Scientific Biography;* N. M. Swerdlow, "Science and Humanism in the Renaissance: Regiomontanus's Oration on the Dignity and Utility of the Mathematical Sciences," *World Changes: Thomas Kuhn and the Nature of Science;* N. M. Swerdlow, "The Recovery of the Exact Sciences of Antiquity: Mathematics, Astronomy, and Geography," in *Rome Reborn;* and Ernst Zinner, *Regiomontanus: His Life and Work.*

13 *Gerard of Cremona's text* The description of the *Almagest* is derived from the useful presentations in Swerdlow and Neugebauer, *Mathematical Astronomy in Copernicus's De Revolutionibus,* Rocky Kolb's *Blind Watchers of the Sky,* Owen Gingerich's "Ptolemy, Copernicus, and Kepler," in *Eye of Heaven,* and G. J. Toomer's entry on Ptolemy, *Dictionary of Scientific Biography.*

15 *"Ptolemy broke sharply"* Rosen, *Scientific Revolution,* p. 28.

15 *"If the Lord Almighty"* Koestler, *Sleepwalkers,* p. 72.

18 *"When my teacher"* Hughes, *Regiomontanus's On Triangles of Every Kind,* p. 27.

19 *"You, who wish to study"* Hughes, *Regiomontanus's On Triangles of Every Kind,* pp. 27–28.

20 *"Quite recently I have made"* Rosen, *Scientific Revolution,* p. 171.

22 *The eclipse occurred* Samuel Eliot Morison, *Admiral of the Ocean Sea,* pp. 653–655.

23 *"For if I am not mistaken"* Pedersen, p. 177.

23–24 *he had said of Trebizond,* Hughes, pp. 15–16.

Chapter Three Childhood

30 *Copernicus's ancestors* Biskup, "Copernicus and His World," pp. 8–9.

31 *Copernicus's forebears* Adamczewski, *The Towns of Copernicus*, p. 14.

33 *Copernicus's family* Gassendi and Thill, *Copernicus*, p. 15.

34 *"Apollo . . . is filled with a passion"* Rheticus, *Narratio prima*, from Rosen, *Three Copernican Treatises*, p. 189.

35 *A fifteenth-century chronicle*r Adamczewski, p. 25.

35 *When Nicolaus was ten* Biskup, p. 12; O'Connor and Robertson, p. 1; Thill, p. 24.

Chapter Four Student Years

41 *The university still has the document* Rosen, "Biography," p. 315.

41 *Until Andreas's death in 1518* Ibid., p. 316.

41 *The University of Kraków* Biskup, "Copernicus and His World," p. 15.

42 *the Copernicus brothers took rooms* Markowski, "The Earliest Unknown Excerpts," p. 8.

42 *"There is in Kraków"* Ibid., p. 6, and Adamczewski, *The Towns of Copernicus*, p. 58.

43 *The early 1490s* Gassendi and Thill, Copernicus, p. 35.

44 *"A comet was visible"* Biskup, "Copernicus and His World," pp. 26–27.

45 *The* Alfonsine Tables Czartoryski, "The Library of Copernicus," pp. 365–67; Rosen, "Biography," p. 326.

46 *"this work is rare"* Rosen, *Scientific Revolution*, p. 172.

46 *Ptolemy stated* Rheticus, *Narratio prima*, p. 111.

47 *Another important work* Westman, "Copernicus, Nicolaus," *Encyclopaedia Britannica;* Thill, pp. 41–42.

48 *he "was not so much"* Rheticus, *Narratio prima,* p. 111.
48 *"he lectured on mathematics,"* Ibid.
49 *Some Copernicus scholars believe* Swerdlow and Neugebauer, *Mathematical Astronomy,* p. 6.

Chapter Five Warmia

52 *Part of that inventory* Swerdlow and Neugebauer, *Mathematical Astronomy,* p. 8.
52 *Written in Latin* Passages and references to Copernicus's *Commentariolus* are based on Rosen's translation, in *Three Copernican Treatises,* pp. 57–90.
56 *"Not only do all"* Copernicus, *On the Revolutions,* p. 11.
57 *the calendar was not reformed* Swerdlow and Neugebauer, p. 8.
58 *"I intend to set forth my views"* Copernicus, "Letter Against Werner," in Rosen, *Three Copernican Treatises,* p. 106.
61 *The canons* Biskup, *Regesta* p. 61.
61 *A visitor in the 1530s* Strauss. *Sixteenth-Century Germany,* p. 63.
62 *"It is a level, plain land"* Ibid.
62 *At mass, they wore vestments* Biskup, *Regesta,* p. 37.
63 *the names of two* Szperkowicz, *Nicolaus Copernicus, 1473–1543,* p. 46.
63 *In addition to collecting taxes* Biskup, *Regesta,* pp. 54–70.
64 *One recipe that survives* Gassendi and Thill, *The Life of Copernicus,* p. 68.
65 *These official minutes* Biskup, *Regesta,* pp. 70–97.
66 *The hostilities ended* Gassendi and Thill, *The Life of Copernicus,* pp. 99–113.

Chapter Six Before the Storm

68 *The other ignominious event* Rosen, "Biography," *Three Copernican Treatises*, p. 319; and O'Connor and Robertson, "Nicolaus Copernicus," p. 4.

70 *Ferber's anti-Protestant zealousness* Biskup, *Regesta*, pp. 143–146.

75 *If that skull is really Copernicus's* Biskup, "Copernicus and His World," p. 10.

75 *"Just as he loved privacy"* Letter from Giese to Donner, in Rosen, *Copernicus and the Scientific Revolution*, p. 165.

76 *"In the year 1501"* Biskup, *Regesta*, pp. 143, 146.

77 *Andreas eventually did leave* Gassendi and Thill, *The Life of Copernicus*, pp. 71–74; Biskup, *Regesta*, p. 93.

78 *Most distressing of all* Biskup, *Regesta*, p. 154.

78 *The Wapowski letter* Ibid., pp. 155–56; Rosen, *Three Copernican Treatises*, pp. 374–75; Swerdlow and Neugebauer, pp. 17–18.

79 *"you maintain that the earth moves"* Rosen, pp. 187–88.

Chapter Seven The Death of the Bishop

81 *"The Chapter of Warmia notifies"* Biskup, *Regesta*, p. 162.

82 *After a night at an inn* Ibid., p. 163.

82 *Bishop Johannes Dantiscus* Ibid., pp. 162–63.

83 *Dantiscus had bragged* Axer and Skolimowska, *Corpus . . . Dantisci*, p. 171.

83 *As Ferber's health declined* Ibid., pp. 209–341.

85 *Dantiscus's main interests* Koestler, *The Sleepwalkers*, p. 180.

86 *"I have received"* Koestler, p. 181; Rosen, *Minor Works*, pp. 323–24.

87 *"Truly, Your Rev. Lordship"* Koestler, p. 182; Rosen, *Minor Works*, p. 326.

88 *Dantiscus later wrote to Giese* Biskup, *Regesta,* pp. 169–70.

88 *Dantiscus even asked Copernicus* Ibid., p. 170.

Chapter Eight The Mistress and the Frombork Wenches

89 *The two aging canons joined Dantiscus* Biskup, *Regesta,* p. 172.

90 *Bishop Dantiscus enjoyed wine.* Axer and Skolimowska, *Corpus . . . Dantisci,* p. 368.

90 *He even mentioned to a correspondent* Ibid., p. 349.

91 *three Danish bishops* Ibid., pp. 157–58, 188.

91 *Personal issues* Ibid., p. 191.

92 *compelled to chide Dantiscus* Ibid., pp. 145–46.

92 *The groom-to-be* Ibid., p. 194.

92 *Dantiscus would not send* Ibid., pp. 194–95, 307, 309–10.

93 *He quoted Saint Paul* Ibid., pp. 191, 256–57.

93 *The woman was named Anna Schilling* Rosen, "Biography," *Three Copernican Treatises,* p. 370; Gassendi and Thill, pp. 232–33.

94 *Anna was reported* Biskup, "Copernicus and His World," p. 44.

94 *"For it must be feared"* Rosen, *Scientific Revolution,* p. 169.

95 *"My lord, Most Reverend Father"* Ibid., p. 151; Rosen, *Minor Works,* pp. 332–33.

96 *Six weeks later* Rosen, *Minor Works,* pp. 334–35.

97 *"I hope that he [Copernicus] will"* Rosen, *Scientific Revolution,* p. 152.

97 *"God Almighty will strengthen"* Ibid., pp. 152–53.

98 *"I am sending back"* Ibid., p. 154.

98 *"As regards the Frombork wenches"* Ibid., pp. 156–57.

99 *"He [Copernicus] is renowned"* Ibid., pp. 158–59.

100 *"Talked earnestly to Doctor Nicholas"* Ibid., p. 160.

Chapter Nine The Taint of Heresy

101 *Tiedeman Giese confirmed* Rosen, *Scientific Revolution*, p. 160.
102 *Scultetus was awarded* Rosen, "Biography," p. 382.
102 *Scultetus's unwarranted scheme* Ibid., p. 383.
103 *"I also hear that my benefice"* Ibid., p. 384.
104 *"With all my might"* Ibid., p. 384.
104 *The first mention* Ibid., p. 384.
105 *Scultetus was thrown in jail* Ibid., pp. 385–86.
106 *From his correspondence,* Axer and Skolimowska, *Corpus . . . Dantisci*, pp. 133–368.
106 *In a formal filing* Rosen, pp. 163–65.
106 *What was Copernicus's attitude* Copernicus, *On the Revolutions*, p. 9.
107 *Giese stated in the preface* Swerdlow and Neugebauer, *Mathematical Astronomy*, pp. 21–22.
107 *"His Reverence has also deemed"* Rosen, pp. 162–63.

Chapter Ten The Catalyst

110 *Luther mentioned that* Manschreck, *Melanchthon*, p. 24.
110 *"Luther rose, extended his hand"* Kesten, p. 272.
112 *The court records are obscure* Danielson, *The First Copernican*, pp. 15–17.
113 *Gasser taught Rheticus* Kraai, p. 89.
115 *"[a]n oddly mixed collection"* Wittenberg Historical Society.
115 *In this disputation* Kraai, p. 18.
116 *To cast proper horoscopes* Ibid., p. 39.
117 *"the science of nativities"* Ibid., pp. 40–41.
118 *Melanchthon was unusually enthusiastic* Ibid., pp. 10, 28; and Manschreck, p. 104.
119 *"It has been reported to us"* Ibid., p. 66.
120 *"To the great Joachim Camerarius"* Ibid., p. 72.

Chapter Eleven The Nuremberg Cabal

123 *Rheticus and Gugler* The description of Nuremberg is based on the superb book by Gerald Strauss, *Nuremberg in the Sixteenth Century*.

125 *Bernard Walther* Donald deB. Beaver, "Bernard Walther: Innovator in Astronomical Observation," p. 39.

126 *the legacy of Regiomontanus* Rosen, "Schöner, Johannes," p. 199.

127 *"When I was with you"* Rheticus, *Narratio prima,* in Rosen, p. 162.

128 *another publisher in Nuremberg* The profile of Johannes Petreius is based on Shipman, "Johannes Petreius, Nuremberg Publisher."

130 *"I heard of the fame"* Rheticus, *Copernicus on Triangles,* in O'Connor and Robertson, p. 5.

Chapter Twelve The Meeting

134 *"Under the penalty of losing head"* Kesten, *Copernicus and His World,* p. 266; Rosen, *Scientific Revolution,* pp. 161–163.

136 *The third and final book* Czartoryski, "The Library of Copernicus," p. 356.

137 *The cathedral courtyard* Biskup, "Copernicus and His World," p. 61.

138 *The "Rulers of Ptolemy"* Copernicus, *Revolutions,* book 4, ch. 15, pp. 232–33; Gassendi and Thill, pp. 118–19.

138 *Tycho Brahe* Gassendi and Thill, p. 123.

138 *A second instrument Copernicus used* Copernicus, *Revolutions,* book II, ch. 2, pp. 67–68.

139 *"I rather compare him with Ptolemy"* Rheticus, *Narratio prima,* p. 109.

Chapter Thirteen The First Summer

140 *"But from the time"* Rheticus, *Narratio*, p. 162.
142 *"[Copernicus] was social by nature"* Ibid., p. 192.
142 *When Rheticus told Copernicus* Ibid., p. 195.
143 *"I see that my teacher always has"* Ibid., p. 163.
143 *Rheticus left a description* Rheticus, Preface to his 1550 *Ephemeridae*, in Kraai, p. 288.
144 *"an amazing fable of Aesop"* Ibid., p. 288.
145 *"I had a slight illness"* Rheticus, *Narratio*, p. 109.
145 *"In his old age"* Rosen, *Scientific Revolution*, pp. 158–59.
146 *"[w]hen Doctor Nicolaus"* Ibid., pp. 160–61.
147 *"He [Giese] realized"* Rheticus, *Narratio*, p. 192.
147 *Rheticus goes on to say* Ibid., p. 192
148 *"the Pythagorean principle"* Ibid., pp. 192–93.
148 *"'The Master says so'"* Ibid., p. 193.
148 *Giese was also realistic* Ibid., p. 195.

Chapter Fourteen Convincing Copernicus

150 *The short book* This and the following quotes are from Rheticus, *Narratio prima*, in Rosen, *Three Copernican Treatises.*
153 *In mid-February 1540* Rosen, "Biography," p. 394.
153 *Osiander wrote back to Rheticus* Rosen, *Scientific Revolution*, p. 192.
154 *"The book may differ"* Danielson, "Achilles Gasser and the Birth of Copernicanism," *JHA*, pp. 460–61.
155 *Petreius meant business* Swerdlow, "Petreius's Letter," pp. 270–74.
156 *On July 1, 1540, he wrote* Biskup, p. 187.
157 *In July 1541* Danielson, *The First Copernican*, pp. 80, 213.
158 *"the shelter of the Muses"* Biskup, *Regesta*, pp. 198–99.
158 *Nicolaus Copernicus revised the last page* Danielson, *The First Copernican*, p. 221.

Chapter Fifteen The Publication

159 *"mention of a certain new astrologer"* Rosen, *Scientific Revolution*, pp. 182–83.
160 *Melanchthon wrote . . . in October 1541* Kraai, p. 117.
160 *The title page identified the author* Biskup, *Regesta*, p. 205.
161 *"no greater human happiness"* Danielson, *The First Copernican*, p. 95.
161 *Erasmus Reinhold* Biskup, p. 204.
161 *"Prussia has given us"* Ibid., p. 206, and Koestler, p. 169.
162 *Leipzig was a fine university* Danielson, p. 105.
163 *"For this art [astronomy]"* Copernicus, *On the Revolutions*, p. 7.
164 *Copernicus suffered a debilitating stroke* Rosen, pp. 165–66.
164 *His condition was confirmed* Biskup, *Regesta*, p. 209.
164 *"On my return"* Rosen, 1984, p. 167.
165 *"Concerning this letter"* Danielson, *The First Copernican*, p. 113.

Chapter Sixteen The Death of Copernicus

170 *"[Copernicus] loved his privacy"* Rosen, *The Scientific Revolution*, pp. 165–66.
171 *Copernicus took his last breath* Ibid., pp. 167–68.
171 *Achacy von Trenck purchased the tower* Biskup, p. 212.
172 *Leonard Niederhoff purchased the* curia Ibid., pp. 212–13.
172 *Copernicus had left a last will* Ibid., pp. 212–13.
172 *the chapter wrote to Dantiscus* Rosen, *Scientific Revolution*, pp. 168–69.
172 *"For it must be feared"* Ibid., p. 169.

Chapter Seventeen Rheticus after Copernicus

175 *"In this way"* Rosen, *Scientific Revolution,* p. 168.
175 *"since the demise of . . . Copernicus"* Danielson, "Achilles Gassen," pp. 457–74
176–177 *not overly impressed with Cardano* Danielson, *The First Copernican,* p. 121.
177 *At one point he was bedridden* Ibid., p. 218.
178 *In April 1551* Ibid., p. 144.
179 *"When I returned"* Koestler, *The Sleepwalkers,* p. 193.
179 *Rheticus died of a respiratory infection* Danielson, *The First Copernican,* p. 193.

Chapter Eighteen The Impact of *On the Revolutions*

181 *Copernicus's work* Gingerich, *The Eye of Heaven,* p. 256
182 *"I know of a recent scholar"* Gingerich, "Reinhold," *Dictionary of Scientific Biography.*
183 *Reinhold was very impressed, though* Gassendi and Thill, p. 254.
183 *One reader of Stadius's new tables* Ibid., 260–61.
187 *"What need is there"* Gassendi and Thill, p. 248.
188 *"When I was studying"* Gingerich, "Kepler," *Dictionary of Scientific Biography,* p. 290.
189 *"God is being celebrated in astronomy"* Ibid., p. 291.
189 *"Kepler established himself"* Ibid., p. 292.
190 *"God let me be bound"* Ibid., p. 295.
190 *"With reasoning I derived"* Ibid., p. 297.
191 *"I accepted Copernicus's theory"* Gassendi and Thill, pp. 285–86.

SELECT SOURCES

Primary Sources

Translated Works of Copernicus

Copernicus, Nicolaus. *On the Revolutions of the Heavenly Spheres (De revolutionibus orbium coelestium libri sex)* (the English title is more fully *Six Books on the Revolutions of the Heavenly Spheres*). Translated by Charles Glenn Wallis. 1939.

The translation appears in Stephen Hawking, ed., *On the Shoulders of Giants: The Great Works of Physics and Astronomy*, pp. 7–388. Philadelphia; Running Press, 2002.

Copernicus, Nicolaus. *On the Revolutions*. Translated by Edward Rosen. In *Nicholas* [sic] *Copernicus: Complete Works*. Vol. 2. Baltimore: Johns Hopkins University Press, 1992.

Letters of Copernicus translated by Edward Rosen

Rosen, Edward. *Copernicus and the Scientific Revolution*. Robert E. Krieger Publishing, Malibar, FL: 1984. Cited: Rosen, *Scientific Revolution.*

Copernicus, Nicolaus. *Nicholas* [sic] *Copernicus: Minor Works.* Baltimore: Johns Hopkins University Press, 1985. Cited: Rosen, *Minor Works.*

Three Copernican Treatises: The Commentariolus of Copernicus, the Letter Against Werner, and the Narratio Prima of Rheticus. 3rd edition. New York: Octagon Press, 1971. Cited: Rosen, *Three Copernican Treatises.*

Dantiscus, Rheticus, and Regiomontanus

Axer, Jerzy, and Anna Skolimowska, eds. *Corpus Epistularium, Ioannis Dantisci, Latin Letters, 1537*. Warsaw: Polish Academy of Arts and Sciences, 2004.

The letters of Johannes Dantiscus dating from 1537. Cited: *Corpus . . . Dantisci.*

The letters of Rheticus were translated into German by, and appear in, Karl Heinz Burmeister, *Georg Joachim Rhetikus, 1514–1574.* 3 vols. Wiesbaden: Guido Pressler, 1966–68.

Regiomontanus, *On Triangles of Every Kind.* Edited by Johann Schöner. Translated by Barnabas Hughes. Madison, WI: University of Wisconsin Press, 1967.

Biographical Sources

Key Biographical treatments of Copernicus cited in this book

Biskup, Marian. *Regesta Copernicana* (Calendar of Copernicus's Papers). Wroclaw, Poland: Polish Academy of Science Press, and in *Studia Copernicana* (Translated by Stanislaw Puppel) 8 (1973).

Gassendi, Pierre, and Olivier Thill. *The Life of Copernicus (1473–1543).* Fairfax, VA: Xulon Press, 2002.

Kesten, Hermann. *Copernicus and His World.* New York: Roy Publishers, 1945.

Koestler, Arthur. *The Sleepwalkers.* London: Hutchinson, 1959.

Rosen, Edward. "Biography of Copernicus." In *Three Copernican Treatises,* pp. 313–408. New York: Octagon Books, 1971, Cited: Rosen, "Biography."

Swerdlow, Noel M., and O. Neugebauer. *Mathematical Astronomy in Copernicus's De Revolutionibus.* New York: Springer-Verlag, 1984. Cited: *Mathematical Astronomy.*

Westman, Robert S. "Copernicus, Nicolaus." *Encyclopaedia Britannica.*

Key biographical treatments of Rheticus cited in this book

Burmeister, Karl Heinz. *Georg Joachim Rhetikus.* 3 vols. Wiesbaden: Guido Pressler, 1967–68.

Danielson, Dennis. *The First Copernican: Georg Joachim Rheticus and the Rise of the Copernican Revolution.* New York: Walker, 2006.

Kraai, Jesse. "Rheticus' Heliocentric Providence: A Study Concerning the Astrology, Astronomy of the Sixteenth Century." Ph.D Diss., University of Heidelberg, 2001.

Important Secondary Sources

Adamczewski, Jan. *The Towns of Copernicus*. Warsaw: Interpress Publishing, 1972.

Applebaum, Wilbur, ed. *Encyclopedia of the Scientific Revolution: from Copernicus to Newton*. New York: Garland 2000.

Barker, Peter. "Reinhold, Erasmus (1511–1553)." *Encyclopedia of the Reformation* (pp. 560–61).

Beaver, Donald deB. "Bernard Walther: Innovator in Astronomical Observation." *Journal of the History of Astronomy* 1 (1970), 39–43.

Biskup, Marian. "Copernicus and His World." *Copernicus: Scholar and Citizen*. Ed. Marian Biskup and Jerzy Dobrzycki. Warsaw: Interpress Publishing, 1972, pp. 7–64.

Bobrick, Benson, *The Fated Sky: Astrology and History*. New York: Simon & Schuster, 2005.

Brachvogal, Eugen. *Frauenburg*. Translated by Reinhard Bechtold. Elbing, Germany: Preussenverlag, 1933.

Burmeister, K. H. "Georg Joachim Rheticus as a Geographer and His Contribution to the First Map of Prussia." *Imago Mundi*, 23 (1969), 73–76.

———."Rheticus, Georg Joachim." *Dictionary of Scientific Biography*, vol. 11, pp. 395–98.

Cardano, Girolamo. *The Book of My Life*. Translated by Jean Stoner. New York: New York Review of Books, 2002.

Czartoryski, Pawel. "The Library of Copernicus." *Studia Copernicana* 16 (1978), 355–96.

Danielson, Dennis. "Achilles Gasser and the Birth of Copernicanism." *Journal of the History of Astronomy* 35 (2004), 457–74.

Dictionary of Scientific Biography. Edited by Charles C. Gillispie. New York: Scribner's, 1970–80.

Drake, Stillman. "Galilei, Galileo." *Dictionary of Scientific Biography*: vol. 5, pp. 237–49.

Ferry, Patrick T. "The Guiding Lights of the University of Wittenberg and the Emergence of Copernican Astronomy." *Concordia Theological Quarterly* 57 (1993), 265–91.

Folkerts, Menso. "Werner, Johannes." *Dictionary of Scientific Biography*, vol. 14, pp. 272–77.

Gingerich, Owen. *The Eye of Heaven: Ptolemy, Copernicus, Kepler*. New York: American Institute of Physics, 1993.

———. *The Book That Nobody Read: Chasing the Revolutions of Nicolaus Copernicus*. New York: Walker and Co., 2004.

———. "Kepler, Johannes." *Dictionary of Scientific Biography*, vol. 7, pp. 289–311.

———, ed. *The Nature of Scientific Discovery*. Washington: Smithsonian Institution Press, 1975.

———. "Reinhold, Erasmus." *Dictionary of Scientific Biography*, vol. 11, pp. 365–67.

Grafton, Anthony. *Cardano's Cosmos: The Worlds and Works of a Renaissance Astrologer*. Cambridge, MA: Harvard University Press, 1999.

Grendler, Paul F. *The Universities of the Italian Renaissance*. Baltimore: Johns Hopkins University Press, 2002.

Hellman, C. Doris, and Noel Swerdlow. "Peurbach, Georg." *Dictionary of Scientific Biography*, vol. 10, pp. 473–79.

Hellman, C. Doris. "Brahe, Tycho." *Dictionary of Scientific Biography*, vol. 2, pp. 401–17.

Hooykaas, R. *G. J. Rheticus' Treatise on Holy Scripture and the Motion of the Earth*. Amsterdam: Elsevier, 1984.

———. "Rheticus' Lost Treatise on Holy Scripture and the Motion of the Earth." *Journal for the History of Astronomy* 15 (1984), 77–80.

Kish, George. "Apian, Peter." *Dictionary of Scientific Biography*, vol. 1, pp. 178–79.

Kostrowiccy, Irena, and Jerzy Kostrowiccy. *Poland*. Warsaw: Arkady, 2004.

Kolb, Rocky. *Blind Watchers of the Sky: The People and Ideas That Shaped Our View of the Universe*. Reading, MA: Helix Books/Addison-Wesley, 1996.

Kuhn, Thomas. *The Copernican Revolution*. Cambridge, MA: Harvard University Press, 1957.

Lemay, Richard. "The Late Medieval Astrological School at Cracow and the Copernican System." *Studia Copernicana* 16 (1978), 337–54.

Lindberg, David C., and Robert S. Westman, eds. *Reappraisals of the Scientific Revolution*. New York: Cambridge University Press, 1990.

Manschreck, Clyde. *Melanchthon: The Quiet Reformer*. New York: Abingdon Press, 1958.

Markowski, Mieczslaw. "The Earliest Unknown Excerpts from Nicolaus Copernicus's *De Revolutionibus*." *Studia Copernicana* (1973), 27–29.

Morison, Samuel Eliot. *Admiral of the Ocean Sea*, Boston: 1942.

Mosley, Adam. "Early Modern Books." Starry Night Website, 1999.

Nunn, George E. "The Lost Globe Gores of Johann Schöner, 1523–1524." *Geographical Review* 17 (1927), 476–80.

O'Connor, J. J., and E. F. Robertson. "Nicolaus Copernicus." The MacTutor History of Mathematics Archive. St. Andrews, Scotland: University of St. Andrews, 2002.

Pedersen, Olaf. "The Decline and Fall of the *Theorica Planetarum*." *Studia Copernicana* 16 (1978), 157–85.

Rabin, Sheila. "Nicolaus Copernicus." *Standford Encyclopedia of Philosophy*. 2005.

Ritvo, Lucille B. "Hartmann, Georg." *Dictionary of Scientific Biography*, vol. 6, pp. 144–45.

Rosen, Edward, and Erna Nilfstein. *Copernicus and His Successors*. London: Hambledon Press, 1995.

Rosen, Edward. "Copernicus, Nicholas." *Dictionary of Scientific Biography*, vol. 3, pp. 401–11.

———. "Novara, Domenico Maria." *Dictionary of Scientific Biography*, vol. 10, pp. 153–54.

———. "Osiander, Andreas." *Dictionary of Scientific Biography*, vol. 10, pp. 245–46.

———. "Regiomontanus, Johannes." *Dictionary of Scientific Biography*, vol. 11, pp. 348–52.

———. "Schöner, Johannes. *Dictionary of Scientific Biography*, vol. 12, pp. 199–200.

Scheible, Heinz. "Melanchthon, Philipp." *Encyclopaedia of the Reformation*, pp. 41–45.

Shank, Michael H. "Regiomontanus, Johannes (1436–1476)." *Encyclopedia of the Scientific Revolution*, p. 560.

———."Peurbach, Georg (1423–1461)." *Encyclopedia of the Scientific Revolution*, p. 492.

Shipman, Joseph C. "Johannes Petreius, Nuremberg Publisher of Scientific Works, 1524–1550." *Homage to a Bookman*. Ed. Lehmann-Haupt. Berlin, 1967.

Stachiewicz, Wanda. *Copernicus and the Changing World.* Montreal: Polish Institute of Arts and Sciences, 1973.

Stachurski, Andrzej. *Warmia and Mazury: Cities and Towns.* Olsztyn: Agencja Fotograficza Mazury, 2006.

Strauss, Gerald. *Nuremberg in the Sixteenth Century.* Bloomington: University of Indiana Press, 1976.

———. *Sixteenth-Century Germany: Its Topography and Topographers.* Madison, WI: University of Wisconsin Press, 1959.

Swerdlow, N. M. "Annals of Scientific Publishing: Johannes Petreius's Letter to Rheticus," *ISIS* 83 (1992), 270–74.

———." Copernicus, Nicolaus (1473–1543)." *Encyclopedia of the Scientific Revolution*, pp. 162–68.

———." The Derivation and First Draft of Copernicus's Planetary Theory: A Translation of the *Commentariolus* with Commentary." *Proceedings of the American Philosophical Society* 1117 (1973), pp. 423–512.

———. "The Recovery of the Exact Sciences of Antiquity: Mathematics, Astronomy, and Geography." In Grafton, *Rome Reborn: The Vatican Library and Renaissance Culture.* Washington, D.C.: The Library of Congress, 1993.

———. "Science and Humanism in the Renaissance: Regiomontanus's Oration on the Dignity and Utility of the Mathematical Sciences." In *World Changes: Thomas Kuhn and the Nature of Science.* Ed. by Paul Horwich, pp. 131–68. Cambridge, MA: MIT Press, 1993.

Szperkowicz, Jerzy. *Nicolaus Copernicus, 1473–1543.* Warsaw: Polish Scientific Publishers, 1972.

Toomer, G. J. "Ptolemy, Claudius." *Dictionary of Scientific Biography,* vol. 11, pp. 187–205.

Treu, Martin, Ralf-Torsten Speler, and Alfred Schellenberger. *Leucorea.* Wittenberg: Editions Hans Fufft, 1999.

Voelkel, James R. "Rheticus, Georg Joachim (1514–1574)." *Encyclopedia of the Scientific Revolution,* pp. 570–71.

Vollman, William T. *Uncentering the Earth: Copernicus and the Revo-*

lutions of the Heavenly Spheres. New York: W. W. Norton & Company, 2006.

Westman, Robert S. "The Melanchthon Circle, Rheticus, and the Wittenberg Interpretation of the Copernican Theory," *ISIS* 66 (1975), 165–93.

Westman, Robert S., ed. "Introduction." *The Copernican Achievement.* Berkeley: University of California Press, 1975.

———"Proof, Poetics, and Patronage: Copernicus's Preface to *De Revolutionibus." The Copernican Achievement.*

———"The Wittenberg Interpretation of the Copernican Theory." *The Copernican Achievement.*

Wilson, Curtis A. "Rheticus, Ravetz, and the 'Necessity' of Copernicus' Innovation." In Westman, *The Copernican Achievement.*

Wrightsman, Bruce. "Andreas Osiander's Contribution to the Copernican Achievement." In Westman, *The Copernican Achievement,* pp. 213–43.

Zinner, Ernst, *Regiomontanus: His Life and Work.* Translated by Ezra Brown. New York: North-Holland, 1990.

Suggested Additional Readings

If you have enjoyed this book, then perhaps you will want to do further reading about Copernicus and the history of astronomy. Here are my suggestions.

If you would like to dig deeper into the science behind the work of Copernicus and the brilliant astronomers who followed him, there are a number of fine books by outstanding historians of science to which you can turn. The single best book, if you are comfortable with mathematics, is Noel Swerdlow and Otto Neugebauer's *Mathematical Astronomy in Copernicus's De Revolutionibus* (1984, Springer-Verlag). In my opinion, this is the single most complete, authoritative source for Copernicus's life and science. In addition, all the books by Edward Rosen, who was the dean of Copernicus scholars for the better part of the twentieth century, are extraordinary useful and interesting. In particular, you should consult *Three Copernican Treatises* (1971, Octagon Press), *Copernicus and the Scientific Revolution* (1984, Krieger Publishing), and *Copernicus and His Successors* (1995, Hambledon Press, London).

For the period immediately after Copernicus, when Reinhold, Brahe, Kepler, and Galileo held center stage, the works of Owen Gingerich and Robert Westman stand out. Gingerich's most relevant papers are collected in *The Eye of Heaven: Ptolemy, Copernicus, Kepler* (1993, American Institute of Physics). Westman's seminal studies can be found in *The Copernican Achievement* (1975, University of California Press) and in David Lindberg and Westman's *Reappraisals of the Scientific Revolution* (1990, Cambridge University Press).

Most of the books cited above are scholarly in nature. They are available in libraries or via online booksellers or auction sites.

For readers looking for lighter fare—that is, books written at about the same level as *Copernicus' Secret*—I have four recommendations.

First, you should read Arthur Koestler's *The Sleepwalkers* (1990, Penguin), a true tour de force. This book covers the history of astronomy from the Greeks to Newton. It was a best-seller when it first appeared in the late 1950s. Every writer wishes he could construct a narrative the way Koestler could. This older title is still easily available as an inexpensive paperback. After Koestler, you should read the more authoritative and equally well-written book by Rocky Kolb, *Blind Watchers of the Sky* (1996, Addison-Wesley), which is the best popular history of astronomy from Tycho Brahe to the twentieth century. With Koestler and Kolb under your belt, I would then recommend two recently published titles. Owen Gingerich's *The Book That Nobody Read* (2004, Walker) is a unique look at the influence that Copernicus's *On the Revolutions* had on the scholarly and astrological communities in the generations immediately following the astronomer's death. And very recently, the first biography of Joachim Rheticus in English appeared, Dennis Danielson's *The First Copernican* (2006, Walker), a terrific and compact study of Copernicus's catalyst.

If you are interested in astrology in the fifteenth and sixteenth centuries, two fascinating books are Anthony Grafton's *Cardona's Cosmos: The Worlds and Works of a Renaissance Astrologer* (1999, Harvard University Press), and Benson Bobrick's *The Fated Sky: Astrology in History* (2005, Simon & Schuster).

Finally, I have enjoyed a clutch of recent books that cover related subjects in the history of astronomy: John Christianson's *On Tycho's Island* (2002, Cambridge University Press), James Connor's *Kepler's Witch* (2005, Harper), Kitty Ferguson's *Tycho and Kepler* (2002, Walker), Timothy Ferris's *Coming of Age in the Milky Way* (2003, Harper), James Gleick's *Isaac Newton* (2004, Vintage), and Dava Sobel's *Galileo's Daughter* (2000, Penguin).

Acknowledgments

As a professional book publisher, I have become a true believer in the importance of the formally published printed word. And, as a publisher of science books, I have developed a deep interest in the seminal ideas that drove the scientific revolution leading to our modern world. These twin passions of mine could not have found a better subject than the publication of Nicolaus Copernicus's *On the Revolutions*. It was the first transcendently important science book published after the invention of the printing press, and the heliocentric theory explained in its pages was the first transcendently important idea of the scientific revolution. When I first learned the unlikely story behind the publication of this justly celebrated title while researching my first book (which was about James Hutton and the discovery of the ancient age of the earth), I could not wait to explore this incredible episode in the history of ideas and hopefully bring it to life.

If I have succeeded, a great deal of the credit goes to my outstanding editor, Bob Bender at Simon & Schuster. He has carefully read each of my successive drafts, and made a number of key recommendations and observations. I now know why he is considered one of the best editors in the business. Everyone at Simon & Schuster has been very professional and helpful, and I have enjoyed working with them a great deal. In particular, I would like to thank Victoria Meyer, Tracey Guest, Johanna Li, Phil Metcalf, and Toby Yuen.

My agent, Susan Rabiner, herself a great science book editor earlier in her career, has also been a wonderful source of ideas, advice, and support. She has my heartfelt appreciation for always being in my corner.

I would like to thank my daughter Christie for timely and careful research assistance, and my wife, Donna, to whom the book is dedicated, for all of her love, support, incredible patience, and excellent proofreading skills! Thanks also to the following friends and colleagues who read the manuscript and offered useful suggestions: Mary Donaldson, Denis Jensen, Mark Ligos, and Wynn Sullivan.

Among the most enjoyable aspects of working on this book were the two trips to Europe that my wife and I took in 2005 and 2006. In 2005, enjoying the hospitality of our good friends the Brauners and the Hoghs in their lovely city of Ulm (and the company of our intrepid parents, Claire, Jack, and MaryLou), we visited the Rheticus sites of Feldkirch, Wittenberg, and Nuremberg. Walking the streets of these beautifully preserved towns helped to transport us back to the sixteenth century. But, the sense of transportation was taken to a level that I can't even describe when we visited Poland in 2006. Gdansk and Torun are stunning, but Frombork and the surrounding towns that Copernicus lived in and often visited are living museums. To stand in Copernicus's front yard—the very ground that he walked on daily—on a gorgeous, soft October day, and to observe the full moon from the quiet streets of Frombork, the huge cathedral on the hill silhouetted by its light, are moments that I will never forget. I am so grateful to the townspeople of Frombork who made our stay so delightful, especially the staff at the Hotel Kopernik. But, in particular, I need to express my gratitude to Elzbieta Topolnicka-Niemcewicz, the curator of the Copernicus Museum in Frombork for her hospitality and significant assistance. Elzbieta runs a wonderful institution in a highly professional way.

Index

Adalbert of Brudzewo, 43, 47
Aeneid (Virgil), 12
Aesop, 129, 144, 151
Age of Exploration, 6
Albert (the Elder), last Teutonic Grand Master and first Duke of Prussia, 154
  Lutheran conversion of, 73, 129
  Warmia's invasion and, 66, 67
Aldebaran, 48
Alderete, Diego Gracian de, 92
Aleksander Jagiellon (Alexander I), King of Poland, 84
Alexandria, Egypt, 13
Alfonsine Tables, 15, 45, 48, 148, 184
Alfonso X, King of Castile, 15
allodiums (granges), 60, 172
Almagest (Ptolemy), 13, 17, 24, 117, 136, 167, 182, 195
  Greek edition of, 136
Alpha Tauri, 7
amber, 27–30, 39
"America," on early globe, 127
Apian, Peter, 121, 127, 135
Apian, Philip, 165–66
Aristarchus, xiv
Aristotle, xiii, xiv, 13, 14, 52, 148, 151, 195
astrolabes, 16, 138, 169
astrology, 4, 5, 6, 7
  astronomy and, 4, 5, 6, 7, 12, 17, 20, 22, 42–43
  Bible and, 37
  Brahe and, 185
  Copernicus and, 36, 42–43, 45, 47, 78, 182, 195
  Kraków University as center for, 42–43
  medicine and, 49
  medieval influence of, 36–37
  Melanchthon and, 115–18, 126, 153, 160
  Nuremberg as center for, 125
  Peurbach and, 12, 17, 127
  Pico della Mirandola's attack on, 47
  Ptolemy's *Almagest* and, 13
  Regiomontanus and, 20, 22, 127
  Rheticus and, 113–14, 115, 116–17, 118, 120–21, 130, 143, 151–52, 160, 176–77
  rigorous (scientific) discipline of, 127
  Schöner and, 126–27, 174
Astronomia nova (Kepler), 190

astronomy:
- *Alfonsine Tables* and, 15, 45, 48, 148, 184
- Arab contributions to, 49
- astrology and, 4, 5, 6, 7, 12, 17, 20, 22, 42–43
- calendars and, 24, 57, 78, 151, 195
- Columbus's voyages and, 44
- Copernicus's interest in, 4–5, 7, 137–38, 142, 147
- Greek language essential to study of, 48
- instruments used in, 5, 137–38, 186
- Italy and, 6–7, 8
- Kraków University as center for, 42–43, 44
- Medieval institutions and promotion of, 4–5, 7
- Peurbach's contributions to, 11–13, 16–18, 144
- Regiomontanus's contributions to, 11–25
- Renaissance and revival of, 6
- Rheticus and, 127

Augsburg, Germany, 123
Augsburg Confession (1530), 110
Augustine, Saint, 129

Baltic coast, 8, 27, 49, 122–23
Baltic Sea, 26, 33, 59, 61, 137
Barczewo, Poland, 89
Basel, Switzerland, 135, 136, 157
benefices, 63, 78, 103, 104
Bessarion, Johannes, 17–19, 79
Bible, 37
- Reformation and, 72, 127

Black Death (bubonic plague), 5–6, 24, 183
blood letting, 20, 64
Bohemia, 20
Boleyn, Anne, 91
Bologna, University of, 6, 8, 41, 45–48, 51, 76
books:
- prices of, 181
- Regiomontanus's "Index" of, 22–23, 127, 155
- standard print runs for, 181
- value and scarcity of, 135–36
- *see also* publishing

Brahe, Tycho, xiv, 138, 184–87, 195
- ambition of, 4
- astrology and, 185
- background of, 184
- benefactors of, 5, 190
- *Commentariolus* named by, 185
- death of, 190
- duelling disfigurement of, 184
- Hven Island observatory of, 185, 187
- Kepler and, 187, 189–90
- observations of, 184–85, 186
- in Prague, 186–87, 190
- Rudolphine Tables of, 190, 191
- Tyconic (geoheliocentric) theory of, 186–87, 190
- at Wittenberg, 184

Braniewo, Poland, 89
bubonic plague (Black Death), 5–6, 24, 183
Buda, Hungary, 20, 44
Buda, University of, 44
Bullinger, Heinrich, 105
Bylica, Martin, 44

calendar, Julian, 24, 57
calendars, 24, 57, 78, 151, 195
Camerarius, Joachim, 129
  Melanchthon and, 120–21, 127
  Rheticus and, 162
Canon of the Science of Triangles, The (Rheticus), 179
Cardano, Girolamo, 129, 176–77
cartography, 42, 44, 102, 125, 126–27, 156–57
Casimir III (The Great), King of Poland, 41
Casimir IV, King of Poland, 43
Cathedral of the Ascension of Our Lady Mary and St. Andrew Apostle (Frombork Cathedral), 7–8, 59–60, 61, 82, 171, 173
celibacy, 58, 93
Chapter of Warmia, 51, 59, 60, 63, 65, 69, 81, 87–88, 94, 96, 100, 104, 107, 172
Charles I, King of Spain, 85
Charles V, Holy Roman Emperor, 72, 85
Chelmno, 59, 82–83, 86, 88, 99, 145
Cologne, Germany, 123
Columbus, Christopher, 6, 22, 43–44, 76
comets, 17, 44, 186
Commentariolus (Copernicus), 8–9, 52–56, 57, 130, 138, 150, 185
"Commentary on Paul's Epistle to the Hebrews" (Bullinger), 105
Congregation of the Index, 194
Conrad of Mazovia, 27
Constantinople, Turkish conquest of, 11, 18
constellations, 14, 38
Copernicus, Andreas:
  as canon, 49, 76, 77
  childhood and youth of, 32–33, 34, 38
  death of, 41, 75–76, 77, 102
  education of, 39, 41, 44, 45, 46, 76
  illness of, 76–77, 85
Copernicus, Nicolaus:
  as administrator of Warmia, 81–82
  ancestors of, 30–32
  Anna Schilling and, 92–100, 101, 109, 142, 145–46
  astrology and, 36, 42–43, 45, 47, 78, 182, 195
  astronomical instruments of, 5, 137–38, 186
  astronomical observations of, 6–7, 48, 51, 65, 67, 125, 137, 151
  astronomy as interest of, 4–5, 7, 137–38, 142, 147
  birth of, 26, 33
  in Bologna, 6, 45–48, 51, 76
  burial of, 171

Copernicus, Nicolaus *(continued)*
as canon, xv, 3, 4, 5, 7–8, 12, 49, 51–52, 58–67, 77–78, 82, 85, 87–88, 101
celibacy rules and, 58, 93
childhood and youth of, 32–38
Columbus's voyages and, 43–44
Commentariolus of, 8–9, 52–56, 57, 130, 138, 150, 185
Dantiscus and, 83, 86–88, 89–100, 101, 103, 109, 145–46, 157–58, 164
death of, 171
demeanor of, 74–75
education of, 3, 6, 7, 36, 39–50
estate of, 171–72
Ferber and, 1–3, 68–70, 74, 77–78, 81–82
in Ferrara, 6, 7, 50, 51
first observation of, 6–7, 48, 51
in Frombork, xv, 2, 7–8, 49, 51–52, 60–65, 67, 69, 75, 77, 85, 92–108, 109, 132–45, 149–59, 164–67
heliocentric theory of, *see Commentariolus*; heliocentric (sun-centered) theory; *On the Revolutions of the Heavenly Spheres*
heresy allegation, 106, 157
in Italy, 6–7, 45–50, 51, 76
at Kraków, 39–45, 51, 52, 58, 84, 184
last will and testament of, 172
"Letter Against Werner" of, 57–58, 126, 130
letters of, 1–3, 57–58, 68, 86–87, 95, 156
library of, 44–45, 48, 135–36, 184
in Lubawa, 145–48
mathematical skills and achievements of, xiv, 7, 51, 55–56, 83, 126, 140, 143, 147–48, 150, 160–61, 168–69, 196
medical studies of, 49
Novara and, 6–7, 47–48, 51, 142
in Olsztyn, 65–67, 69
in Padua, 6, 49, 51
personality of, 75, 102, 141–42
Peurbach and Regiomontanus as precursors of, 11–25
as physician, 64–65, 69, 81, 88, 97, 101, 134, 145, 157
planetary tables compiled by, 78, 139
posthumous reputation of, 181–96
priesthood refused by, 68–69
Prussian map project of, 102, 156–57
Ptolemaic system doubted by, 15, 54, 56
Ptolemy compared with, 139, 140, 150, 161
recognition of work of, 78–80
Reformation and, 106–7

Regiomontanus and, 20, 139, 140, 150
Rheticus and manuscript of, 138–48, 156–58, 172, 190, 195
Rheticus's first meeting with, 134–36
Rheticus's journey to meet with, xv, 130–31, 132–34
in Rome, 48, 51
Schönberg's letter to, 79–80, 83, 108, 168
scientific method of, 142–44
scientific revolution and, xiii–xv, 3–5
scripture and theory of, 175, 188, 195
Scultetus and, 101–2, 146, 156–57
stroke and paralysis of, 164, 167, 170–71
Teutonic Knights invasion and, 65–67, 69, 70
uniform circular motion as tenet of, xiv, 54, 56, 183, 190, 195
Watzenrode as benefactor of, 7, 35–36, 39, 41, 59, 75, 76
as Watzenrode's secretary, 4, 7, 51, 69, 75, 82, 83
Copernicus, Nicolaus (father), 31–32, 35, 75
Copernicus, Nicolaus (grandfather), 31
Cosmic Mystery, The (*Mysterium cosmographicum*) (Kepler), 188–89, 191
Cosmographia (Ptolemy), 23–24
Counter-Reformation, 103–6, 107, 134, 135, 142, 189
Crusades, 27
curias, 60–61, 75, 93, 96, 134, 137, 149, 171–72

Dantisca, Juana, 85, 92, 93
Dantiscus, Johannes, 82–88
ambassadorial career of, 84–85, 92
Anna Schilling affair and, 92–100, 101, 109, 145–46, 172–73
benefice issue and, 103, 104
as bishop of Chelmno, 82–83
as canon, 83, 85–86
Copernicus and, 83, 86–88, 89–100, 101, 103, 109, 145–46, 157–58, 164
Counter-Reformation and, 107, 134, 142
death of, 173
education of, 84
finances of, 91–92
Luther and, 110–11
Melanchthon and, 110
Scultetus's conflict with, 101–6, 142
Warmia bishopric voted to, 87–88
Warmia tour of, 89–90, 93, 103, 109
Darwin, Charles, 181
De Astronomia Libri IX (Geber), 135
deferents, 14–15, 54
Delgada, Isabel, 85, 92
De mundi aetherei recentioribus phaenomenis (Brahe), 186–87

Denmark, 85
  Reformation in, 73, 91
Dialogue Concerning the Two Chief World Systems, The (Galileo), 193–94
Disputations against Divine Astrology (Pico della Mirandola), 47
Dobre Miasto, Poland, 89
doctrine of first motion, 140–41
Donner, George, 172
"Do the Laws Condemn Astrological Prognostications?" (Rheticus), 115
Ducal Prussia, *see* Prussia
Dürer, Albrecht, 110

earth, xiii, xvi, 13, 36–37, 161, 188
  Brahe's Tyconic system and, 186–87
  Copernicus's axiom about, 9, 54, 55
  "eccentric" and "equant" points of, 15, 54, 183
  inclination of axis of, 169
  moon as satellite of, 9, 13, 55, 194
  rotation of, 9, 47–48, 54, 55, 153, 169, 194
  see also geocentric (earth-centered) (Ptolemaic) theory
Easter, 24, 47
eccentric, 15
eclipses, 7, 16, 17, 20, 22, 44, 48, 141
Eighth Sphere, 58
Einstein, Albert, xiv
Elbe River, 111, 137
Elblag, Poland, 1, 2, 32, 59
  Reformation in, 70, 74
Elements (Euclid), 45, 135, 136
Elijah, 152
England, 85
  Reformation and, 91
Ephemerides (Regiomontanus), 22, 24
epicycles, 14–15, 54
Epitome Astronomiae Copernicanae (Kepler), 191
Epitome of the Almagest (Regiomontanus and Peurbach), 17–19, 20, 24, 46, 47, 136, 190
equant, 15, 54, 183
equinoxes, 24
Erasmus, Desiderius, 129
Euclid, 23, 45, 135, 136, 177
Europe:
  amber trade in, 27–30, 39
  Black Death (bubonic plague) in, 5–6, 24, 183
  central, map of, *40*
  Counter-Reformation in, 103–6, 107, 134, 135, 142, 189
  publishing industry in, *see* publishing
  Reformation in, xv, 70–74, 91, 93, 100, 103–12, 114–15, 117–20, 126, 129–30, 163
  Regiomontanus as first scientific publisher in, 21–25, 26
  Renaissance beginning in, 6
  trade in, 27–30, 33–34, 39
  universities in, 6–8, 10,

11–13, 16, 18, 39–50; *see also specific universities*

Feldkirch, Austria, 177
Gasser in, 113–14, 154, 161
Rheticus in, 112–14, 161
Ferber, Maurice, 86, 88, 89
as canon in Frombork, 69
Copernicus and, 1–3, 68–70, 74, 77–78, 81–82
Dantiscus and, 92
death of, 81–82, 83
illness of, 68, 81, 83
Reformation and, 70, 73, 74, 106
Scultetus and, 102–3
Ferrara, University of, 6, 7, 8, 50, 51
"first orders," 3, 58, 69
fixed stars, *see* stars
Flachsbinder, Johannes, *see* Dantiscus, Johannes
Foucault, Jean-Bernard-Léon, 194
Francis I, King of France, 91
Frederick III, Holy Roman Emperor, 17
Frisches Haff (Vistula Bay), 61, 132, 137
Frisius, Gemma, 85, 158, 164, 183
Frombork, Poland, 12, 59–65, 73, 83, 85–87
Andreas Copernicus as canon in, 49, 76, 77
Anna Schilling's banishment from, 94, 172–73
burning of, 66
Cathedral Hill in, 59–60, 61, 96, 137, 185
cathedral in, 7–8, 59–60, 61, 82, 171, 173
Copernicus in, xv, 2, 7–8, 49, 51–52, 60–65, 67, 69, 75, 77, 85, 92–108, 109, 132–45, 149–59, 164, 167
Copernicus's *allodium* (grange) in, 172
Copernicus's *curia* (house) in, 60–61, 75, 93, 96, 134, 137, 149, 171–72
descriptions of, 62, 132–33
Ferber in, 69
map of, 53
Morsing's stay at, 185–86
Rheticus in, 134–45, 149, 156–59
St. Nicolaus church in, 61

Galileo Galilei, xiv, 191–94, 195
astronomical discoveries of, 192–93
background of, 191
Catholic Church and, 193–94
house arrest of, 194
Kepler and, 189, 191, 193
at Padua, 192–93
at Pisa, 191–92, 193
religious views of, 193
telescope and, 192, 193
Gama, Vasco da, 6
Gasser, Achilles, 113–14, 115, 127, 129, 154, 155, 157, 161
Copernicus book publication supported by, 154, 155, 157
Narratio prima and, 154, 155, 157

Gasser, Achilles *(continued)*
new edition of Copernicus book urged by, 175–76
Gdansk, Poland, 8, 26, 32, 34, 39, 59, 69, 94, 102, 149, 170
as Baltic trading center, 122–23
Reformation and, 73–74, 91
Geber, 135
geocentric (earth-centered) (Ptolemaic) theory:
Arab astronomers' skepticism of, 49, 56
Aristotle and, xiii, xiv, 13, 14, 52, 148, 151, 195
astronomical observations as disproofs of, 184–85, 186, 192, 193
Brahe's merging of heliocentric and, 186–87, 190
Catholic Church and, xiii
Copernicus's skepticism of, 8–9, 46
Kepler's advocacy of heliocentric theory over, 188–89
lunar motion in, 46
Novara's skepticism of, 7
planetary motion in, 13–15, 47
pre-Copernican doubters of, xiii–xiv
Ptolemy and, xiii, xiv, 13–16, 22, 46, 52, 54, 56, 152–53, 183
Regiomontanus's criticism of, 46
retrograde motion in, 14, 15
geography, 43
geometry, 135, 177, 179, 188
Geometry (Euclid), 177
Gerard of Cremona, 13, 17, 136
Germany, 26, 41, 42, 111
Jews in, 124
public school system in, 126
publishing industry in, 21–25, 26, 126, 127–29, 135–36, 153–56, 166, 174
Reformation in, 72, 73, 109–12, 114–15, 117–20, 126, 129–30, 163
Giese, Tiedemann, 75, 83–84
Anna Schilling affair and, 99–100, 101, 145–46
as bishop of Warmia, 173
Chelmno bishopric voted to, 88
Copernicus book publication and, 145–48, 154, 164, 166–67
Copernicus's death and, 171
Copernicus's manuscript and, 172
Copernicus's stroke and, 164, 170
Counter-Reformation and, 107
death of, 173
illness of, 145
Narratio prima and, 154
new edition of Copernicus book urged by, 175
Scultetus and, 103
Goethe, Johann Wolfgang von, ix
granges (*allodiums*), 60, 172
gravity, 9, 188, 194, 195
Graz, Austria, 187, 189
Greek language, 48

Gugler, Nicholas:
in Nuremberg, 123–25
Rheticus's Nuremberg trip with, 122–23
Gutenberg, Johann, 11, 128
gymnasiums, 126

Halley's comet, 17
Hanseatic League, 26, 33, 39
Hartmann, Georg, 127, 160
heliocentric (sun-centered) theory, xvi
Brahe's merging of geocentric and, 186–87, 190
Church's suppression of, 193–94
Commentariolus and, 8–9, 52–56, 57, 130, 138, 150, 185
Copernicus's concern for the correctness and completeness of, xiv, 10, 80
Copernicus's reluctance to publish his full work on, xiv, 9–10, 80, 108, 146, 147
Copernicus's seven axioms for, 8–9, 54–56
Copernicus's technical details for support of, xiii, xiv
early Church interest in, 78–80
Galileo and, 191–94
Kepler's advocacy of, 188–89
Luther's dismissal of, 159
Melanchthon's rejection of, 160
Narratio prima as primer on, 149–50, 152–53
Newton and, 188, 194, 195
Reinhold's disregard of, 182–83
scripture and, 175, 187, 188–89, 193–94
see also On the Revolutions of the Heavenly Spheres (Copernicus)
Henry VI, Holy Roman Emperor, 27
Henry VIII, King of England, 85, 91, 129
Hen War, 91
heresy, 104–5, 106, 135, 157
"higher orders" (Church vows), 3, 59, 69
Hipparchus, 151
Holy League, 111
Holy Roman Empire, 17, 18, 27, 85, 112
Reformation and, 72, 74
horoscopes, 20, 37, 49, 116, 117, 151, 185
Hosius, Stanislaw, 103–4, 135, 142
Hungary, 20, 41, 44
Hven Island, Denmark, 185, 187

Imperial pill, 64–65
"Index of Books" (Regiomontanus), 22–23, 127, 155
Index of Forbidden Books, 194
indulgences, 71
Innocent VIII, Pope, 44
Inquisition, 193–94
Instrumentum primi mobilis (Apian), 135
Iserin, Georg, 112–13

Islam, Muslims, 152
Italy, 17, 41
  astronomy community in, 6–7, 8
  Copernicus in, 6–7, 45–50, 51, 76
  publishing in, 128–29

Jeriorany, Poland, 89
Jesuits (Society of Jesus), 104
Julian calendar, 24, 57
Jupiter, 13, 55, 186, 194
  Galileo's discovery of moons of, 193

Kepler, Johannes, xiv, 187–91, 195
  astrological calendar of, 187–88
  benefactors of, 5
  Brahe and, 187, 189–90
  death of, 191
  education of, 187, 188
  elliptical orbit of planets discovered by, 190–91
  as one of the first true Copernicans, 188, 190–91
  Galileo and, 189, 191, 193
  in Graz, 187, 189
  in Prague, 189–90
  religious views of, 187, 188–89, 190
  Rudolphine Tables and, 190, 191
  scripture and Copernican theory viewed as harmonic whole by, 188–89
  as teacher, 187
Knights Templar, 27
Kopernik, Mikolaj, *see* Copernicus, Nicolaus
Koperniki, Poland, 30
Kraków, Poland, 30–31
  Copernicus in, 39–45, 51, 52, 58, 84, 184
  Rheticus in, 178–79
  Wawel Castle in, 39
Kraków, Treaty of (1525), 73
Kraków, University of, 8, 39–45, 51, 52, 58, 84, 85, 156, 184
Królewiec, Poland, 1, 2, 26, 68, 73, 157
Kunheim, George von, 157

Ladislaus V, King of Hungary, 17
Leipzig, University of, 16, 162, 165, 174, 176, 177
  Rheticus's banishment from, 178
Leipzig Disputation (Luther), 71
Lemnius, Simon, 120–21
Leonardo da Vinci, 45
  Last Supper of, 6
Leo X, Pope, 57
  Luther's excommunication by, 71–72
leprosy, 76–77
"Letter Against Werner, The" (Copernicus), 57–58, 126, 130
Letters on Sunspots (Galileo), 193
Lidzbark, Poland, 45, 51, 73, 81–82, 85, 88, 89, 96
Lindau, Switzerland, 177
Lithuania, Kingdom of, 27
lodis, 39

Loitz, Michael, 172
Lubawa, Poland, 86, 87, 98, 99
  Copernicus's visit to Giese in, 145–48
Lubeck, Germany, 123
lunar eclipses, 7, 17, 22, 44, 48, 141
Luther, Martin, 70–72, 107, 118, 129, 135, 166
  Dantiscus and, 110–11
  death of, 173
  excommunication of, 71–72
  heliocentric theory and, 159
  Leipzig Disputation of, 71
  Lemnius and, 120
  Ninety-five Theses of, 71, 114
  Osiander and, 130
  personality of, 109, 119
  as publisher, 127–28
  Rheticus's horoscope of, 116
  Wittenberg home of, 114–15
  world chronology of, 152
Lutherans, Lutheranism, xv, 70–74, 91, 93, 126, 129–30, 134, 135, 142
  Catholic sacraments and, 105
  Nuremberg as center of, 72, 73, 126, 129–30
  Scultetus's sympathy toward, 104–5, 135, 142, 157
  Wittenberg as heart of, 72, 109–12, 114–15, 119
  see also Reformation, Protestant

Maestlin, Michael, 187, 188
Mainz, Germany, 11, 128
Marlbork, 59
Mars, 13, 55, 186, 189
mathematics:
  Cardano and, 129
  Copernicus and, xiv, 7, 51, 55–56, 83, 126, 140, 143, 147–48, 150, 160–61, 168–69, 196
  Galileo and, 191, 192
  Peurbach and, 16, 19
  Regiomontanus and, 19, 20
  Rheticus and, 113, 116, 135, 138–39, 142, 160–61, 178–80
  Werner and, 125
Matthew of Miechow, 43, 52
Matthias Corvinus, King of Hungary, 19–20
Maximilian I, Holy Roman Emperor, 85
medicine, 20, 49, 64–65
Melanchthon, Philipp, 109–21, 129, 135, 166, 184
  appearance of, 110
  astrology as interest of, 115–18, 126, 153, 160
  Augsburg Confession and, 110
  Camerarius and, 120–21, 127
  death of, 173
  Dürer's portrait of, 110
  German public school system and, 126
  Gugler and, 122
  Lemnius scandal and, 120–21
  Narratio prima and, 153
  Osiander and, 130
  as publisher, 127–28
  Rheticus's appointment as professor by, 115–16

Melanchthon, Philipp *(continued)*
  Rheticus's departure from Wittenberg approved by, 162
  Rheticus's horoscope of, 116
  Rheticus's meeting with, 109, 111–12
  Wittenberg University and, 109–12, 114, 115–16, 117–19, 153, 160, 162, 181, 182
Mercury, 13, 47, 55, 56, 186
meteorology, 43
Meusel, Hans, 178
Michelangelo, 6, 45
Milan, Italy, 129, 176
Milichius, Jacobus, 115–16, 136
Moldavia, 91
Montulmo, Antonius de, 155
moon, xiii, 17, 20, 37, 141, 169, 186
  Aldebaran's occultation by, 48
  Copernicus's axiom about, 9, 55
  phases of, 47
  in Ptolemaic system, motion of, 46
  as satellite of earth, 9, 13, 55, 194
Morsing, Elias Olsen, 185–86
Müller, Johannes, *see* Regiomontanus

Narratio prima (First Report) (Rheticus), 141, 147, 160, 166, 167, 171, 178
  as advance publicity for Copernicus's work, 154–55, 157–58
  Commentariolus compared with, 150
  description of, 149–53
  heliocentric theory and, 149–50, 152–53
  Osiander's enthusiasm about, 153–54
  publication of, 149
  second edition of, 157
navigation, 44
Netherlands, 85
New Testament, vernacular German edition of, 72
New Theory of the Planets (Peurbach), 16–17, 22, 26, 161, 182
Newton, Isaac, xiv, 188, 194, 195
  Kepler and, 191
New World, 6, 76, 127
Niederhoff, Leonard:
  as cantor of Frombork chapter, 88, 106
  cohabitation issue and, 93, 97, 99
  as Copernicus executor, 172
  Copernicus's *curia* owned by, 171–72
  heresy allegation, 106, 157
Ninety-five Theses (Luther), 71, 114
North Star, 38
Nostrodamus, 183
Novara, Domenico Maria de, 58
  astrological almanac of, 47, 48
  Copernicus and, 6–7, 47–48, 51, 142
Nuremberg, Germany, 21–24, 26, 133, 142
  astrology and, 125

astronomical observatory at, 21, 25
description of, 123–25
Jews expelled from, 124
Kaiserburg Castle in, 123
Pegnitz River in, 124
population of, 123
as publishing center, 126, 127–29, 153–56, 166, 174
Reformation in, 72, 73, 126, 129–30, 163
Regiomontanus in, 21–24, 26, 125, 126
Rheticus in, 112, 123–31, 161, 162
Rheticus's journey to, 122–23
St. Lorenz and St. Sebald churches in, 124

occultation, 48
Old Testament, vernacular German edition of, 72
Olsztyn, Poland, 63, 65–67, 69, 73
On Natal Horoscopes (Montulmo), 155
"On the Dignity of Astrology" (Melanchthon), 118
On the Origin of Species (Darwin), 181
On the Revolutions of the Heavenly Spheres (Copernicus), 21, 48, 57, 129, 138
appeals for new edition of, 175–76
Church banning of, 194
Commentariolus as reference to, 56, 138
complexity of, 58
Copernicus's deathbed viewing of copy of, 171
Copernicus's failure to acknowledge Rheticus in, 166–67, 168
Copernicus's final corrections for, 156–58, 162
Copernicus's "Preface and Dedication" in, 168
Copernicus's reluctance about publication of, xiv, 9–10, 80, 108, 146, 147
delay in publication of, xiv–xv
effort to secure Copernicus's publication of, 79–80, 83, 108, 149–56, 168
impact of, 181–96
Kepler and, 188
original manuscript of, 156, 172
Osiander's anonymous preface to, 163–64, 165, 166, 168, 175, 183
preparation and publication of, 159–68
Reinhold and, 161, 181–83, 187
Rheticus and, 138–58, 160–61, 162, 164, 165–66, 167, 168, 174, 175, 177, 178–79, 182, 190
Rheticus's fair copy of, 156, 158, 160, 161
Schönberg's letter to Copernicus in, 168
scientific revolution's beginning with publication of, xiii–xiv

On the Revolutions of the Heavenly Spheres (Copernicus) *(continued)*
second edition of, 178–79
structure of, 140–41, 168–69
Wapowski and, 139
On the Sides and Angles of Triangles (Copernicus), 160–61, 166, 167
On Triangles of Every Kind (*De triangulis omnimodis*) (Regiomontanus), 19, 24, 127, 135, 136, 160
Opus palatinum de triangulis (Rheticus), 180
orbits, circularity of, xiv, 54, 56, 183, 190, 195
Orneta, Poland, 89
Osiander, Andreas:
Albert's Lutheran conversion and, 73, 129
Copernicus's book and, 153–54, 155, 162–64, 165, 166, 168, 175, 183
Copernicus's correspondence with, 156, 162–63
Luther and, 130
Melanchthon and, 130
personality of, 129–30, 163
Petreius as publisher for, 155–56
Reformation and, 73, 129–30, 163
Rheticus and, 129–30, 153–54, 174
Otto, Valentin, 179–80, 190
Ottoman Empire, 11, 18, 91, 152, 188
Padua, University of, 8, 41
Copernicus at, 6, 49, 51
Galileo at, 192–93
Paracelsus, 113
Paris, University of, 41
"Paul's Epistle to the Romans" (Bullinger), 105
Peasants' War, 72, 111, 126
Pegnitz River, 124
pendulum, 194
Perspectiva (Witelo), 135–36
Petreius, Johannes, 128–29, 135–36, 174, 177
book dedication to Rheticus by, 154–55
Copernicus's book and, 154–56, 160, 161, 162, 164, 165–66, 175, 181
death of, 173
Narratio prima and, 154
Peucer, Caspar, 184
Peurbach, Georg, 11–18, 58, 178, 182, 190
astrology and, 12, 17, 127
Bessarion and, 17–18
death of, 18
as instrument-maker, 16
as mathematician, 16, 19
Regiomontanus's partnership with, 16–18, 25
as self-taught astronomer, 12
as teacher, 11, 12, 16
writings of, 16–19, 20, 22, 24, 26, 45, 46, 47, 136, 144, 161, 182, 190
Pico della Mirandola, Giovanni, 47
Pisa, University of, 191–92, 193
planets, 13–15, 17, 47

Brahe's Tyconic system and, 186–87, 190
celestial latitudes of, 141
Copernicus's description of motions of, 141, 153, 169
Copernicus's ordering of, 54, 55, 56
"deferents" and "epicycles" of, 14–15, 54
gravity and motions of, 194
Kepler's discovery of elliptical orbits of, 190–91
Kepler's emphasis on sun's role in positions of, 188
retrograde motion of, 14, 15
as wandering stars, 13–14, 15–16, 37, 45
see also specific planets
Plotkowski, Paul, 88, 98–99
brother of, 106
Counter-Reformation and, 107
Poland, 27, 30, 43
Counter-Reformation in, 103–6, 107, 134, 135, 142
Jesuits and, 104
Thirteen Years' War and, 31–32, 59, 65
trade in, 33–34, 39
Warmia's alliance with, 73
see also specific cities and rulers
Pomerania, 59
Posnan, Poland, 132
Prague, Bohemia:
Brahe in, 187, 189–90
Rheticus in, 178
Prague, University of, 41
printing, 11, 21–24, 46, 128
see also publishing
Protestant League, 111
Prussia, 26–27, *28*, 52, 73, 74
first map of, 102, 156–57
Prussian Council, 32, 59
Prussian Estates (Royal Prussia), *28*, 31–32, 59, 83, 84
Prussians, 27
Prutenic Tables (Reinhold), 182, 183
Ptolemy, Claudius, xiii, xiv, 13–16, 22, 46, 52, 54, 56, 151, 188
Almagest of, 13, 17, 24, 117, 136, 167, 182, 195
Copernicus compared with, 139, 140, 150, 161
see also geocentric (earth-centered) (Ptolemaic) theory
publishing:
Gutenberg and rise of, 11, 128
Nuremberg as center of, 21–25, 26, 126, 127–29, 153–56, 166, 174
Petreius and, 128–29, 135–36, 154–56, 160, 161, 162, 164, 165–66, 173, 174, 175, 177, 181
Regiomontanus and, 21–25, 26
of scientific works, 21–25, 128, 181
Wittenberg as center of, 127–28

quadrants, 138

Reformation, Protestant, xv, 70–74, 91, 93, 100, 103–12, 114–15, 117–20, 126, 127, 129–30, 163
Regiomontanus, 5, 36, 43, 47, 48, 56, 58, 129, 151, 178, 190
  as astrologer, 20, 22, 127
  astronomical observations of, 20–21
  Bessarion and, 18–19
  calendar reform and, 24, 57
  Copernicus compared with, 139, 140, 150
  death of, 23, 24, 25
  education of, 16
  given name of, 11
  in Hungary, 19–20, 44
  "Index of Books" of, 22–23, 127, 155
  as mathematician, 19, 20
  in Nuremberg, 21–24, 26, 125, 126
  personality of, 23–24
  Peurbach's partnership with, 16–18, 25
  Ptolemaic system challenged by, 46
  as publisher, 21–25, 26
  in Rome, 18–19, 24, 25, 133
  writings of, 18–19, 20, 22–23, 24, 25, 45, 46, 47, 127, 135, 136, 160, 190
Rehden, Dietrich von, 172
Reich, Felix, 82, 88, 107, 170
  Anna Schilling affair and, 92, 93, 96–98
  Dantiscus's Warmia tour and, 89–90, 93, 103
  death of, 98
Reinhold, Erasmus, 184, 187
  death of, 183
  Prutenic Tables of, 182, 183
  Wittenberg University and, 116, 161, 162, 181–83, 184
Renaissance, 51
  beginning of, 6
  Columbus and, 6, 22, 43–44, 76
  Peurbach and Regiomontanus and, 11–25
  printing and, 11, 21–24, 46
Reszel, Poland, 89
retrograde motion, 14, 15
Rheticus, Georg Joachim:
  almanac of, 177
  astrology and, 113–14, 115, 116–17, 118, 120–21, 130, 143, 151–52, 160, 176–77
  astronomy and, 127
  breakdown of, 177
  Camerarius and, 162
  in Cassovia, 179
  Copernicus's book and, 138–58, 160–61, 162, 164, 165–66, 167, 168, 174, 175, 177, 178–79, 182, 190, 195
  Copernicus's final days and, 170, 171
  Copernicus's first meeting with, 134–36
  Copernicus's manuscript and, 138–48, 156–58, 172, 190
  Copernicus's *On the Sides*

and Angles of Triangles published by, 160–61, 166, 167
death of, 179
education of, 113
execution of father of, 112–13
in Feldkirch, 112–14, 161
in Frombork, 134–45, 149, 156–59
Gasser and, 113–14, 115, 127, 154, 155, 157, 161, 175–76
in Gdansk, 149
harmony of Copernican theory and scripture supported by, 175, 188
journey to Frombork by, xv, 130–31, 132–34
in Kraków, 178–79
in Leipzig, 162, 165, 174, 176, 177
Lemnius scandal and, 119–21
in Lubawa, 145–48
as mathematician, 113, 116, 135, 138–39, 142, 160–61, 178–80
Melanchthon and, 109–21, 162, 166
in Milan, 176–77
name change of, 113
Narratio prima of, 141, 147, 149–55, 157–58, 160, 166, 167, 171, 178
in Nuremberg, 112, 123–31, 161, 162
Osiander and, 129–30, 153–54, 174
Otto and, 179–80, 190
Petreius's book dedication to, 154–55
post-Copernican career of, 174–80, 184
in Prague, 178
as professor, xv, 115–16, 117, 119, 127, 130, 135, 157, 159–62, 174, 176, 177
Prussian map project and, 156–57
publishing industry and, 127–28
rape conviction of, 178
trip to Nuremberg of, 122–23
as Wittenberg's dean of faculty, 160
in Zurich, 113
Rhodes, Franciscus, 149
Roman Catholic Church:
benefices of clerics in, 63, 78, 103, 104
calendar reform and, 24, 57
celibacy rules of, 58, 93
Copernican theory and, 79–80, 193–94, 195
Copernicus as canon in, xv, 3, 4, 5, 7–8, 12, 49, 58–67, 77–78, 82, 85, 87–88, 101
"first orders" in, 3, 58, 69
Galileo tried by, 193–94
geocentric theory and, xiii, 52
"higher orders" in, 3, 59, 69
Hosius as Counter-Reformation leader in, 103–4, 135, 142
Index of Forbidden Books of, 194
indulgences and, 71

Roman Catholic Church (continued)
  Reformation and, xv, 70–74, 91, 93, 100, 103–8
  Sacramentarian sect of, 105
  Watzenrode as bishop of, 4, 7, 35, 42, 45, 51, 59, 66, 69, 82, 83
  Watzenrode as canon of, 35, 42
Rome, Italy:
  Andreas Copernicus's probable death in, 77
  Chapter of Warmia and, 59
  Copernicus in, 6–7, 45–50, 51
  Regiomontanus in, 18–19, 24, 25, 133
  St. Peter's Basilica in, 71
  Scultetus in, 102, 105, 170
Rostock, University of, 184
Royal Library of Hungary, 20
Royal Prussia (Prussian Estates), *28*, 31–32, 59, 83, 84
Rudolph II, King of Bohemia, 190
Rudolphine Tables (Brahe), 190, 191

Sacramentarians, 105
Saint Johann school, 36
St. Lorenz Church, 124
St. Peter's Basilica, 71
St. Sebald Church, 124
Salza, Hermann von, 27
Sambia, 32
Saturn, 13, 55, 186
Saxony, 111
Scandinavia, 41
Scepperus, Cornelius, 157–58
Schilling, Anna, 92–100, 101, 109, 142, 145–46
  background of, 93–94
  banishment from Frombork of, 94, 172–73
  in Gdansk, 170
Schmalkaldic League, 74
Schönberg, Nicholas, letter to Copernicus from, 79–80, 83, 108, 168
Schöner, Johann, 112, 121, 122, 129, 132, 160, 161
  astrology and, 126–27, 174
  background of, 126
  as cartographer and globe maker, 126–27
  Copernicus's book and, 155
  death of, 173
  Narratio prima and, 150–51, 154
  as publisher and editor, 126, 127–28, 135
  skills and interests of, 126–27
Schultze, Alexander, *see* Scultetus, Alexander
scientific revolution:
  Copernicus and founding of, xiii–xv, 3–5
Scultetus, Alexander, 88, 97, 98
  attempt at excommunication of, 102–3
  banishment of, 105
  as cartographer, 102, 156–57
  Copernicus and, 101–2, 146, 156–57
  Dantiscus's conflict with, 101–6, 142
  heresy charges against, 104–5, 135, 142, 157

imprisonment of, 105
mistress and children of, 93, 99–100, 102–3, 105
personality of, 101, 102
in Rome, 102, 105, 170
Sigismund I, King of Poland, 73–74, 83, 84, 86, 87–88, 91
Counter-Reformation and, 106
Scultetus and, 102–3, 104, 105, 142
Silesia, 30
Simplicio, 193
sinecures, 63
Sixtus IV, Pope, 24
Snopek, Paul, 88
Society of Jesus (Jesuits), 104
solar eclipses, 17, 44
solstices, 17
Spain, 17, 85, 92, 93
spring equinox, 24
Stadius, Johannes, 183
Starry Messenger, The (*Sidereus Nuncius*) (Galileo), 193
stars, xiii, 14, 37, 38, 45, 55, 58, 151, 161, 169, 186
doctrine of first motion of, 140–41
"Eighth Sphere" of, 58
planets as "wandering" types of, 13–14, 15–16, 37, 45
Suleiman I, Sultan of the Ottoman Empire, 91
sun, xiii, xvi, 17, 141, 151, 153, 169, 186, 188
Copernicus's axiom about, 9, 54, 55, 56
Galileo's observations of, 193
see also heliocentric (sun-centered) theory
sundials, 16
sunspots, 193
supernovas, 184–85, 186, 192
surveying, 19, 43, 44
Sweden, 26
Reformation and, 72
Switzerland, Reformation in, 72
Syntaxis (Greek edition of *Almagest*) (Ptolemy), 136
syphilis, 76

Tables of Directions, The (Regiomontanus), 20, 24, 45, 136
Tables of Eclipses (*Tabulae eclipsium*) (Peurbach), 17, 45, 144
Tatars, 84
Tczew, Poland, 102
telescopes, 5, 13, 192, 193
Teutonic Knights (Order of the Knights of the Cross), 27–30, *28*, 31, 32, 33, 35, 59, 90
Warmia's invasion by, 65–67, 69, 70, 73
Teutonic Order State of Prussia, 27, 129
Theodoric of Reden, 79
Theophrastus, Philippus (Paracelsus), 113
Theory of the Planets, A (*Theorica planetarium*) (Gerard), 12–13, 17
Thirteen Years' War, 31–32, 59
Toruń, Peace of (1466), 32, 65, 66
Toruń, Poland, 31–32, 37, 39, 59, 76, 133
castle in, 35

Toruń, Poland *(continued)*
Copernicus homes in, 33, 34, 35
Copernicus's birth in, 26, 33
descriptions of, 34–35
founding of, 33
Kraków compared with, 39
as trading center, 33–34
trade, 27–30, 33–34, 39
Trebizond, George, 24
Trenck, Achacy von, 88, 146
Trent, Council of (1560), 104
triangulation, 19
trigonometry, 16, 19, 20, 125, 135, 160, 166, 177, 178, 179, 180
triquetrums, 137–38, 186
Tubingen, Academy of, 120
Tübingen, University of, 187, 188, 189
Tyconic (geoheliocentric) theory, 186–87, 190

Ulm, Germany, 123
universe:
Copernicus's *Commentariolus* and conception of, 8–9, 52–56, 57, 130, 138, 150, 185
Copernicus's manuscript's general description of, 140
geocentric theory of, *see* geocentric (earth-centered) (Ptolemaic) theory
heliocentric theory of, *see* heliocentric (sun-centered) theory
Kepler's model of, 188–89
Newton's theory of gravity and understanding of, 188, 194, 195
Ptolemy's *Almagest* as explication of, 13, 17, 24, 117, 136, 167, 182, 195
as sphere, 169
Urban VIII, Pope, 193

Venus, 13, 47, 55, 186
vernal equinox, 24
Vienna, Austria, 18
Vienna, University of, 10, 11–13, 16, 18, 43, 178
Vistula Bay (Frisches Haff), 61, 132, 137
Vistula River, 31, 33, 34, 39
Volmer, Johannes, 115–16

Wallachians, 84
Walther, Bernard, 125
wandering stars, 13–14, 15–16, 37, 45
Wapowski, Bernard, 42, 58, 78, 126
Wapowski, Piotr, 42
Warmia, 8, *28*, 32, 49, 51–67, 81–88
bishop and canons as effective rulers of, 63
chapter of canons of, *see* Chapter of Warmia
Copernicus as administrator of, 81–82
Counter-Reformation in, 103–6, 107, 134, 135, 142
Dantiscus's election to bishopric of, 87–88
Dantiscus's formal tour of, 89–90, 93, 103, 109

Giese as bishop of, 173
Poland's alliance with, 73
Reformation and, 70, 73–74, 91, 93, 103–8
Teutonic Knights attack on, 65–67, 69, 70, 73
Watzenrode as bishop of, 4, 7, 35, 45, 51, 59, 66, 69, 82, 83
Wartburg, Germany, 72
Watzenrode, Barbara, 31
Watzenrode, Katherine Molibog, 75
Watzenrode, Lucas, 31–32, 33, 46, 49
as bishop of Warmia, 4, 7, 35, 45, 51, 59, 66, 69, 82, 83
as canon, 35, 42
Copernicus as secretary of, 4, 7, 51, 69, 75, 82, 83
as Copernicus's benefactor, 7, 35–36, 39, 41, 59, 75, 76
death of, 75
in Lidzbark, 45, 51
Werner, Johannes, 58, 125–26, 130
Widmanstetter, Johann, 79
Witelo, 135–36
Wittenberg, Germany, 67, 133, 134
Castle Church in, 71, 114, 122
layout of, 114–15
Lemnius epigrams published in, 120
Luther's home in, 114–15
as publishing center, 127–28
Reformation and, 72, 109–12, 114–15, 119
Wittenberg, University of:
Brahe at, 184
Luther at, 70–71, 119, 159
Melanchthon as rector of, 109–12, 114, 115–16, 117–19, 153, 160, 162, 181, 182
Peucer as rector of, 184
professorial salaries at, 181
Reinhold and, 116, 161, 162, 181–83, 184
Rheticus and, xv, 109, 115–16, 117, 119, 127, 130, 135, 153, 157, 159–62, 174, 176
Wolsey, Thomas, 85
Worms, Diet of, 71–72

Zell, Heinrich, 132, 134, 149, 156–57, 159
zodiac, 14, 20, 151
Zwingli, Ulrich, 70, 72, 113, 129

VERMONT STATE COLLEGES
0 0003 0837812 4